Discovering ECOLOGY

Patrick H. Armstrong

Shire Publications Ltd.

CONTENTS

PHOTOGRAPHS

Photographs are acknowledged as follows: Aerofilms, plates 10, 12, 13; E. A. Armstrong, plate 3; R. D. Barrett-Lennard, plate 11; the Forestry Commission, plates 1, 5, 6, 7, 8, 9; Eric Hosking, plate 2. Plate 4 is by the author.

INTRODUCTION

The scope of ecology

Ecology is the study of animals and plants in the places where they live—or, to put it a little more scientifically, the investigation of the relationships between organisms and their environment.

It has been customary for ecologists to concentrate on 'natural' environments or habitats (e.g. woods, sea-shores, mountains), but man has had such a massive impact on so wide a range of environments that, certainly in Britain and probably over a large proportion of the earth's surface, truly natural habitats, unaltered by man, are rare. It seems arbitrary, therefore, to make a distinction between woodlands, which have almost certainly been subjected to considerable management, on the one hand, and more obviously artificial grasslands and hedgerows on the other. In this little book the alterations in habitats due to man's activities will be described, together with the inter-relationships that exist between the animals and plants normally to be found in these habitats.

Sometimes a distinction is made between animal ecology and plant ecology. But in a freshwater pool or a field of wheat, animals—small types such as insects, as well as the larger, more conspicuous mammals, birds and amphibians—are constantly using the plant life for shelter and food, and so to study both together is often most rewarding. Again, the subject may be divided into autecology, the detailed study of the ecology of a single type of plant or animal, and synecology, the investigation of the whole community of plants and animals. The latter approach will be adopted here; after a chapter in which the concepts and language of ecology are explained, there follow sections discussing a number of common British habitats.

Why ecology is important

Man depends on his environment for his food, raw materials and water. Thus his food comes from the farmlands and oceans of the world; timber for construction from the forests; water from lakes, rivers and underground sources. Man also affects his surroundings by burning coal, oil and gas and by the production of sewage, refuse and industrial waste. The ecologist hopes that by discovering how animal and plant communities function mankind will be enabled to continue to utilise forests, fisheries and farmlands without over-exploiting them or otherwise damaging them so that their yield is reduced. Conservation and pollution are discussed in the last three chapters.

Scientific names

All organisms, animals and plants, are given scientific or 'Latin' names. (They are in fact sometimes derived from Latin, sometimes from Greek, sometimes just invented.) In this way scientists of different nationalities can immediately recognise what is being discussed. In this book plants and animals will be referred to by their English names where they have them, but usually the first time an organism is mentioned its everyday name will be followed by the scientific one, thus: buttercup *(Ranunculus acris)*, house sparrow *(Passer domesticus)*.

1. THE SCIENCE OF ECOLOGY

Adaptation

Plants and animals have to be adapted to their environment and mode of life in order to survive. Many animals that live in water are streamlined in shape—fish, whales and dolphins, aquatic birds such as penguins and a range of invertebrates (animals without backbones) like the squid *(Loligo forbesi)*— and most of the larger aquatic creatures have fins or flippers for swimming. Other adaptations are the shapes of birds' beaks—often the structure is clearly related to way of life: a bird of prey has a hooked beak for tearing flesh, a kingfisher a long pointed bill for catching fish, finches have short stumpy beaks for eating seeds, and so on.

Adaptations may be behavioural as well as structural; the mole's burrowing ability is as much part of its adaptation to an underground way of life as its strong, shovel-like limbs (see also plate 3).

Cacti and other plants that live in deserts (xerophytes), as well as having a waxy covering preventing excessive water-loss, have fleshy internal tissues that enable them to absorb large quantities of water following a sudden desert storm.

An organism is adapted, therefore, to perform a particular rôle in a community; it occupies a certain position in the total framework. An animal feeds on a restricted range of foods

4

and in its turn is preyed upon by a limited number of predators. An organism may be very restricted in the type of environment it can occupy; some types of fish only live in caves, other creatures only in hot springs; some plants live only in very salty places by the sea. An organism's place in the total scheme of things is its ecological niche.

Habitat factors

Organisms are subjected to a number of environmental influences.

Climatic factors include amount of sunlight, humidity and temperature; organisms vary in their range of tolerance with respect to all of these. Subtle variations in microclimate may exist within very short distances and thus cause the range of animals and plants found to be widely different within a few inches. Woodlice, for instance, lose moisture very quickly and therefore tend to live in humid environments, especially during the day, and are found beneath stones and logs; the little woodlouse *Philoscia muscorum* very soon dies in dry conditions. The damper, north-facing side of a tree-trunk sometimes has a more extensive growth of mosses and algae (the green 'mould' sometimes found on the bark) than the part of the trunk exposed to the sun.

Soil or edaphic factors also have a profound influence on animals and plants; some plants, for example, can tolerate very acid soils with no lime in them; such plants are calcifuges e.g. heather or ling (*Calluna vulgaris*). Others, found on lime-rich soils, are known as calcicoles. Lime is also required by snails for their shells.

The presence or absence of salt in soil (or water) may be a limiting factor. Some plants, such as sea-lavender (*Limonium vulgare*) can survive in very salty soils, while most land plants are killed by salt. Some fish species, e.g. the eel (*Anguilla anguilla*) and the salmon (*Salmo salar*), move from fresh to salt water and back again but most marine or fresh-water creatures have a limited tolerance range and may perish if the salt concentration gets too low or too high.

The biotic factors that affect an organism are those that stem from the activities of other plants and animals in its immediate environment. Grazing by rabbits (*Oryctolagus cuniculus*) on heathland, for example, results in the suppression of species such as heather; very heavy grazing eventually leads to the elimination of almost all plants except the sand sedge (*Carex arenaria*) and a few mosses and lichens. Man, too, may be an important biotic influence on a community; heathland may be burnt, grassland may be grazed and trampled by stock, a pool may be polluted by industrial effluent.

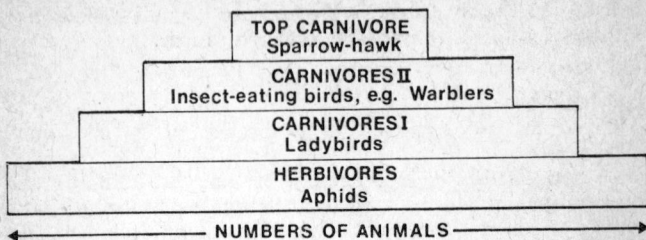

Fig. 1: Pyramid of numbers of animals in English woodland.

Food relationships

Green plants are able to obtain their food from inorganic (i.e. non-living) sources—the atmosphere and water and salts in the soil—by a process known as photosynthesis which uses the energy in sunlight; their green chlorophyll enables them to do this. Animals have to consume either plants or animals that have themselves eaten plants, for they are not able to build up complex organic substances (e.g. fats, proteins) from inorganic materials. Energy and matter are therefore transferred along food-chains (see plate 2). For example, many species of flies obtain their food from plant material; they may be eaten by spiders which in their turn may form part of the diet of small birds such as warblers and titmice. These may be taken by birds of prey. Mice and voles eat acorns and provide an important source of food for stoats. Organisms may thus be classified as producers—the green plants, herbivores—the plant eaters, and carnivores—which obtain the bulk of their food from animal sources. Sometimes the term 'omnivore' is used for animals such as man that obtain their food from a wide range of sources.

Usually, the closer an animal is to the end of a food-chain, the larger it is—thus a stoat is larger than a vole and ladybirds are larger than the tiny insects on which they live. Predatory animals, however, are usually much less abundant than the creatures they consume; thus a pyramid of numbers exists (fig. 1).

One species of organism may provide food for many other species. Oak-trees, for example, support hundreds of species of insects (see page 22). Bank voles *(Clethrionomys glareolus)* may be eaten by stoats, weasels, hawks, owls and a number of other mammals and birds of prey (see plate 2). Food-chains thus branch and join together to form food-webs (fig. 2).

Food relationships other than the producer-herbivore-carnivore food-chain type exist; dead organisms provide food for saprophytic plants. Fungi (the major group of plants to which mushrooms, toadstools and moulds belong) live on the decaying remains of vegetation or animal carcasses. The birdsnest orchid *(Neottia nidus-avis)* is an example of a flowering plant that has a similar mode of nutrition, living off the decaying leaves on the floor of beech-woods. Many organisms obtain their food from living plants and animals; these are parasites. Plant examples include the flowering plants dodder *(Cuscuta epithymum)*, a parasite of ling and one or two other heathland plants, the broomrapes *(Oro-banche)*, which raid the roots of other flowers, and *Pythium*, the fungus that causes 'damping off' of seedlings. All these have no chlorophyll and so are incapable of photosynthesis, but some plants, such as mistletoe *(Viscum album)*, do possess

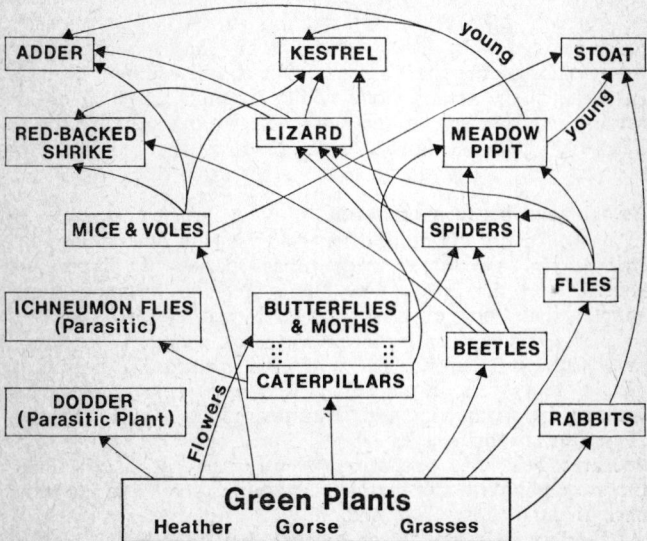

Fig. 2: Simplified food-web for lowland heath.

the green chlorophyll, obtaining only some of the nutrients they need from the living tissues of their hosts; these are partial parasites. Animal examples include ichneumon flies, whose larvae suck the tissues of caterpillars, and the worms that live in the guts of many creatures, as well as lice and fleas. Sometimes the term parasitic is used to describe the cuckoo *(Cuculus canorus)* which lays its eggs in the nests of other species of birds, the young cuckoo later being fed by the host pair.

An essential difference between symbiosis and parasitism is that, although an intimate association exists between the two organisms concerned, neither is damaged, and often both species obtain benefits. Lichens, for example, consist of an alga that lives symbiotically in the company of a fungus. The fungus benefits from the food photosynthesised by its algal partner, the alga obtaining protection from desiccation during dry periods by being surrounded by the fungus. Some algae also live in corals.

A less intimate relationship is known as commensalism. In a strict sense this is a food-sharing association, although very often one party is the provider, getting little in return. The pea-crab *(Pinnotheres pisum)* lives within the shell of the mussel *(Mytilus edulis)*, sharing its food. The pilot-fish *(Naucrates ductor)* swims ahead of a shark, feeding from the remains of the shark's meals. But the best-known example is, perhaps, the sea-anemone which lives on the shell of the hermit crab, feeding on the fragments from the crab's meals. The crab gains at least a measure of protection from the stinging tentacles of the anemone (see pages 47-48).

Associations between organisms

In fact organisms quite often gain benefits other than food from their associations with other species of plants and animals (see also plate 1). There are, for instance, many animals that ' groom ' other species by eating their parasites. In Africa starling-like birds known as ox-peckers *(Buphagus)* perch constantly on the backs of large mammals (rhinoceros, giraffe, hartebeest, cattle, etc.) removing ticks and giving warning by their cries and disturbed flight of the approach of danger. Many species of fish, some shrimps and at least one crab behave in a comparable way, the larger fish sometimes queuing in substantial numbers to be ' cleaned ' and have their parasites removed.

Other animals are ' hitch-hikers ', having become adapted to attach themselves to others for transport; some barnacles live only on the bodies of whales, and sucker-fish have large oval discs on their heads which enable them to become attached to larger fish and turtles. When the carrier catches food the passenger releases hold and shares in the meal. In some tropical regions the fisher-folk attach sucker-fish to long lines and release them to catch sharks and turtles.

Some defenceless creatures take advantage of the protection obtainable by associating with well-armed partners. A number of fish are able to survive swimming amongst the stinging tentacles of jellyfish; others live in a similar association with large tropical species of sea-anemone. Some small species of tropical fish regularly shelter among the long needle-like spines that cover the surface of sea-urchins.

On a much less intimate level, some birds make use of the protection offered by rabbit holes as nesting sites (see pages 11 and 47).

Many plants require the co-operation of insects to achieve fertilisation; minute grains of pollen have to be transferred from the male parts of one flower, the stamens, to the female part, the stigma, of another. Many flowers, therefore, have scents, possess bright colours and secrete nectar, and thus attract insects such as bees and butterflies and so ensure pollination. Sometimes the structure of a flower is adapted to the characteristics of a particular pollinator. Snapdragons (*Antirrhinum majus*) and white dead-nettles (*Lamium album*) are pollinated by bumble-bees; when a bee alights on the lower part of the dead-nettle flower and pushes towards the nectaries (the nectar-producing organs), the stamens bend and strike the insect on the back, smearing it with pollen which the bee may later transfer to another flower. The bee orchid (*Ophrys apifera*) is shaped and coloured like a female bee and when a male bee attempts to mate with it pollen is smeared on its head.

The whole complex of relationships, normal food linkages and other associations, established between the animals and plants in any community is known as a species-network.

Ecosystems

An ecosystem may be defined as a 'segment of nature', including all the animals and plants within it—the whole species-network—plus the inorganic environment in which they live. An ecosystem, therefore, includes the plants and animals, the soils, water and atmosphere, and perhaps also the manifestations of human activities in an area. It may

9

be considered as a community of organisms together with their physical environment. An ecosystem may be of any size—a drop of pond-water, a rock pool by the sea-shore, a copse, an oakwood, an area of heath, or an island.

Ecologists today often try to study whole ecosystems, investigating both the species-network (the organism-organism interactions) and environment-organism relationships (e.g. the effects of some of the habitat factors on the animals and plants).

The sum total of all the ecosystems in the world, the whole part of the earth's crust, its surface waters and atmosphere which provide a medium for life, is sometimes called the biosphere, itself really just a single very large ecosystem.

Competition and the balance of nature

An oak-tree produces thousands of acorns each year, the bank vole has several litters each season with four or five young in each, and a clutch of tawny owl's *(Strix aluco)* eggs may number as many as seven. Yet the world is not overrun by these organisms. This is because far more young animals and plants are produced than can possibly survive to adulthood. Intense competition exists between individuals of the same species—competition for food, water, light, space, breeding partners and so on—and numbers are held in check by predation, disease and the scarcity of food resources.

A given area is able to support a limited number of each of the species found within it. A certain number of voles, stoats, deer and so on can survive in a ten-acre wood and no more. If the vole population temporarily increases beyond a certain point, food supplies become inadequate and a proportion of the animals starve or die from disease. Also, the number of predators tends to increase and before long the number of voles is reduced. Although quite spectacular short-term fluctuations occur, in the long term a state of balance or equilibrium is usually maintained. Sometimes this is referred to as the 'balance of nature'.

Frequently this balance is disturbed by man. A gamekeeper may be particularly zealous in the destruction of 'vermin', although some of the birds and beasts of prey that he kills are the natural enemies of the mice and voles that feed on the beech-mast or acorns that would otherwise grow into saplings. The reduction in numbers of predators thus results in an increase in rodents and the growth of the saplings that might eventually replace older trees is prevented.

The natural woodland regeneration processes are thus disturbed.

Ecological change

Changes in ecosystems, gradual or radical, are constantly occurring.

One of the most spectacular followed the spread of myxomatosis, the rabbit disease, in the 1950s. Infection was reported for the first time in Britain from Kent in October 1953. It spread rapidly through most of lowland England and much of Wales in the summer of 1954 and by the end of 1955 the disease was established in almost every area of England, Wales and Scotland. The effect on the rabbit population was immediate and catastrophic; on one Suffolk farm some five thousand rabbits were shot the season before the spread of the myxomatosis and only thirty-five the following season.

The effect on the plant communities was striking; many species of plants, particularly in grassland and heathy areas, were able to grow up and flower where they had previously been very heavily grazed. In some places bushes and small trees became established.

Various mammals were also affected; the stoat *(Mustela erminea)*, dependent on the rabbit for much of its food, decreased in many areas. On the other hand, hares *(Lepus europaeus)*, which compete with rabbits to some extent for their food, increased. The annual kill of hares on the estate mentioned above increased from an average of 213 per season for the ten years prior to myxomatosis to an average figure of 692 for the 10 years following the outbreak.

There were also many indirect effects; food-chains were disrupted and predator pressure on ground-nesting birds increased. Many of them, for example, the wheatear *(Oenanthe oenanthe)*, formerly quite a common open-ground species, suffered in other ways as well; patches of heavily grazed grass needed as feeding areas disappeared, and so did rabbit-burrows, used as nesting sites by these birds.

The impact of a sudden ecological change may be transmitted throughout the species-network very quickly.

E. P. Odum, in his book *Fundamentals of Ecology*, defines ecological succession as 'the orderly process of community change; it is the sequence of communities which replace one another in a given area.' The whole set of communities in such a series constitutes a sere, and the end-member of the series, the mature community, is called the climax.

A 'hydrosere' is the name given to the set of communities

that develop in wet environments—e.g. a mineral-working that has recently been abandoned, or perhaps the small backwater of a river, cut off from the main stream by silting. The first stage in this type of succession (the pioneer stage) will probably include a number of water-weeds such as water starwort *(Callitriche palustris)*. Gradually the pool will be infilled by the accumulation of dead plant material and by silting and it will become smaller in area and shallower. Around the edge, where the water is very shallow, conditions soon become unsuitable for the growth of truly aquatic species, and reeds *(Phragmites communis)* invade. The build-up of organic material accelerates and eventually the reedswamp is succeeded by a 'carr' community—a low, tangled woodland of willow *(Salix)* and alder *(Alnus)*. This, after several further decades may develop into a damp oakwood (see plate 4).

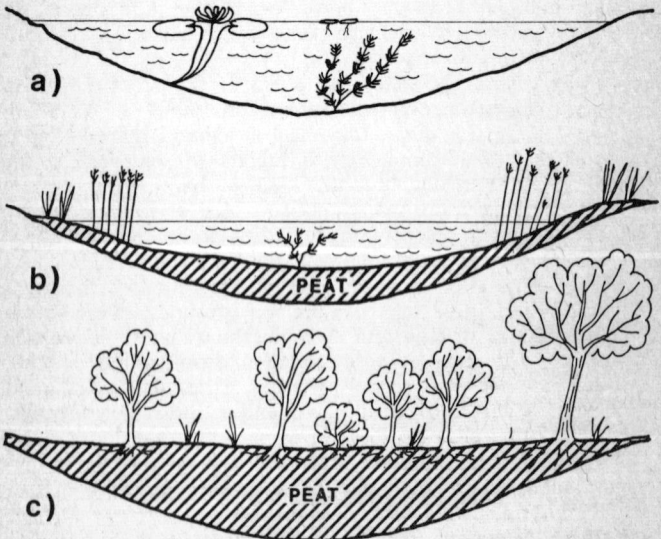

Fig. 3: Stages in a hydrosere—the gradual infilling of a pool: a, open water with aquatic plants such as water-lilies, water starwort, etc.; b, invasion by reedswamp as pool becomes shallower through peat accumulation; c, depression completely filled by peat, development of 'carr' and woodland vegetation.

A psammasere develops in a sand-dune environment; almost the only plant able to live in the unconsolidated sand is marram-grass *(Ammophila arenaria)*. This grows in tussocks amidst the dunes, the roots binding the sand together and the organic material produced enriching the developing soil. After twenty or thirty years the soil may be sufficiently consolidated for other plants to establish themselves. Among the first to follow the pioneer marram-grass is usually ling, and the dune ridge is gradually converted into a type of heathland which, in the absence of grazing or burning, eventually becomes birch or pine scrub.

At Studland in Dorset and between Formby and Southport in Lancashire extensive systems of dunes have accumulated parallel to the sea-shore. In the Dorset area the study of old maps reveals that the innermost of the four dune ridges is over three centuries old and it now supports heath vegetation; the youngest, close to the sea, has formed in the last couple of decades and bears just a few marram spikes. At both Studland and Formby waterlogged depressions or 'slacks' are to be found between the dune ridges. These also, of course, differ in age and so contain contrasting vegetation types. In places stretches of open water exist, e.g. Little Sea, Studland, but elsewhere the slacks have been completely infilled and support a dense willow carr. Dunes are liable to be blown landwards by the wind and near Formby and also on the Culbin Sands, Morayshire, they have been stabilised by establishing conifer plantations.

Animal communities show parallel successions to those of plants; quite different assemblages of birds and insects live, for example, associated with the open water of a pool, in a reedswamp or in woodland. But the succession of animal communities is perhaps best illustrated in a microhabitat, a small-scale environment such as a decaying log or a piece of carrion (i.e. an animal carcass).

In an animal corpse, the larvae of burying beetles *(Necrophorus),* blow-flies, and whole series of other carrion flies and beetles may be at work, the composition of the fauna varying somewhat with the locality and the type of carcass. Eventually the body of an animal ends up as little more than a pile of bones in which other beetles such as *Dermestes* may be found. Ultimately the bones disintegrate into the soil.

The succession in a dead branch may start with the arrival of bark-beetles and longicorn beetles. There are many different species of bark-beetles, each type specialising in the wood of a particular tree; they excavate breeding corridors immediately under the bark. Longicorn beetles may excavate similar galleries or go straight into the wood. While the beetles and their larvae are active, fungi may also attack the wood and eventually the bark will be loosened. Once this has happened, many species will start to colonise the log, eating both the decaying wood and the fungi. The fauna of a quite small fragment of wood at this stage may include several dozen species—woodlice, millipedes, beetles and earwigs. There may also be some organisms, spiders, for example, that prey on the other colonists.

Rather similar successional stages may be identified in dung microhabitats.

The cycling of elements

When organisms die, the material of which they are composed disintegrates and is used again, or is 'recycled'.

Plants take in carbon dioxide during photosynthesis, and while they may pass some of it back to the atmosphere in respiration—the burning up of organic material to provide energy—much is stored in the plants' tissues in the form of carbohydrates (complex chemicals containing carbon dioxide and water). Some of this carbon-containing material is passed along food-chains, the plants being eaten by animals which are then eaten by other animals, but eventually the carbon dioxide is released once more on respiration or following the death of an organism. Bacteria and fungi cause the complex chemicals of which the plants and animals are built to break down and the carbon dioxide returns to the atmosphere.

Under certain conditions, however, for example in water-logged environments, from which oxygen is excluded, the bacteria responsible for decomposition cannot operate, and decay is impeded. In such circumstances peat accumulates. In the geological past peat formed in this way was compressed and ultimately altered to produce coal. Oil and natural gas probably formed from the partial decomposition of animal and plant material that accumulated at the bottom of former seas. Such 'fossil fuels' represent large stores of carbon in the earth's crust—stores which are passed back to the atmosphere on the combustion of the fuels. There is evidence that the amount of carbon dioxide in the atmosphere is gradually increasing.

Other chemical elements circulate in a similar way. The principal reservoir of nitrogen in the environment is in the atmosphere—some 79 per cent of the air is nitrogen—but the majority of living things cannot use it directly. It has first to be 'fixed' by specialised organisms, electrical discharges in the atmosphere or by man in certain industrial processes.

The nitrogen-fixers include certain blue-green algae and a number of types of bacteria. Some of these live free in the soil, but others form symbiotic associations with green plants, particularly leguminous plants. Thus nitrogen-fixing bacteria live in nodules (small swellings) on the roots of plants such as lucerne and clover—these plants are often included in crop-rotations in order to improve the nitrogen status of agricultural soils. Lightning may also oxidise some of the nitrogen in the atmosphere, so some fixed nitrogen falls to the ground dissolved in rainwater during thunderstorms.

The nitrates (salts containing nitrogen) formed by these processes are taken in by the roots of green plants and are used to build up complex proteins, and the plants eventually

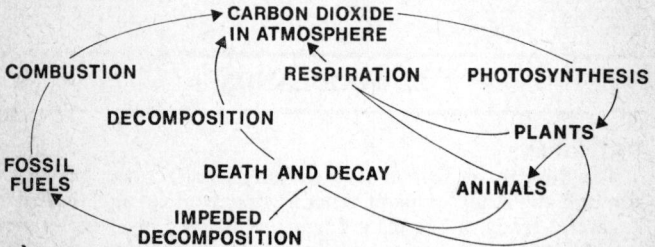

a)

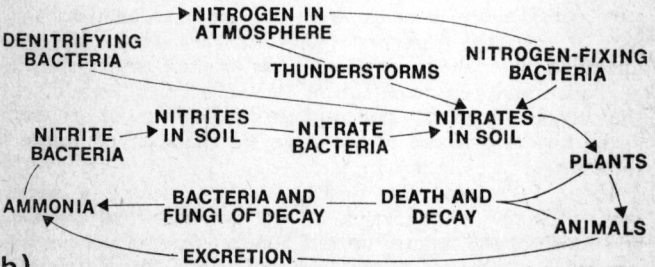

b)

Fig. 4: Simplified diagrams of: a, carbon cycle; b, nitrogen cycle.

15

are eaten by animals. When an organism dies, a series of different types of bacteria break down the proteins, and eventually the nitrogen is returned to the soil. There are also some bacteria, the denitrifiers, that pass the nitrogen from the soil to the atmosphere. Today large amounts of nitrogen are fixed artificially and used as nitrogenous fertilisers.

Phosphorus, in the form of calcium phosphate, is an important constituent of bone; animals obtain it from plant foods and when they die the bones disintegrate into the soil and this element, too, is thus recirculated.

The circulation of oxygen and hydrogen is extremely complex and will not be considered here. Similar principles apply.

Plants and animals require small amounts of 'trace elements'; minute quantities of molybdenum and cobalt appear to be necessary for some plant processes (including the fixation of nitrogen). Copper and boron also seem to be needed by other plants and animals in small quantities and are sometimes added to farmlands. These elements, too, circulate in a broadly similar way to carbon, nitrogen and phosphorus.

2. WOODLANDS

Early history

The history of Britain's vegetation is ascertained primarily through the study of plant remains preserved in peat deposits. A careful study of the plant fragments in successive layers can reveal the changes that have occurred in the vegetation of the area surrounding the site. Pollen grains, though small, vary considerably in shape from one plant to another; they may be extracted from peat (sometimes also from clay) and identified under the microscope. Trees produce large quantities of pollen and thus comparison of the relative amounts of the various types of pollen-grain in deposits of different ages provides evidence of the changing character of Britain's forests.

About 5,000 B.C., the British Isles experienced a warm, moist climate. The glaciers of the Ice Age had retreated and much of the country up to 2,500 feet (c. 800 metres), was covered by a forest of oak, elm and elder. Birch was also present in some of the uplands but was much rarer in the east and south. The Highlands of Scotland carried a cover of

Scots pine *(Pinus sylvestris* var. *scotica)*. In these forests roamed animals such as the bear *(Ursus arctos)* and wolf *(Canis lupus)* and small bands of Mesolithic (Middle Stone Age) men who lived by hunting, fishing and collecting berries.

When, about 3,000 B.C., at the start of the Neolithic period (New Stone Age), the first farmers brought stock-keeping to Britain, almost continuous forest still covered these islands and grassland vegetation was rare. Probably almost the only food available for sheep and cattle was the foliage from trees —elm leaves are particularly nutritious. At about this time the pollen record shows a sharp fall in the amounts of elm pollen being deposited, and it has been suggested that this 'elm decline' was due to the utilisation of the elm on a substantial scale for feeding stock. Another theory is that the decline was due to some infection akin to Dutch elm disease, a fungoid parasite of foreign origin spread by bark beetles and present in Britain since the 1930s.

The clearance of the forests

As is still the case in parts of the tropics today, Neolithic man seems to have cleared patches of forest, planted crops for a few years and then, as yields began to decline, abandoned the clearings. Usually reinvasion by woodland followed, but occasionally, as in the Breckland heaths of East Anglia, the clearance was more lasting (see page 29).

The Bronze Age (beginning about 1,700 B.C.) marked an acceleration of the rate of disforestation, for men were now equipped with metal tools. It was at about this time that the chalk hills of eastern and southern England were cleared. They remained important arable areas throughout the Iron Age and Roman occupation. During the Anglo-Saxon period considerable forest clearance and expansion of agriculture took place on the clay lands of the Midlands and the Weald, perhaps because of the introduction of the eight-ox plough. Meanwhile, on some of the chalk ridges farmland was being abandoned. It is possible that at this time the beech *(Fagus sylvatica)*, previously unable to compete with the oak, became established, forming woodlands in areas such as the Chilterns and the Cotswolds.

The Domesday survey shows that in 1086 forest still covered a much greater portion of England than it does today. In Norman times large areas such as the New Forest in Hampshire and the Forests of Arden and Feckenham in the Midlands were designated royal forests. These lands may not have been entirely covered in woodland; the term 'forest' signified that special laws applied, designed to preserve the king's hunting—particularly wild-boar and deer. Norman woodland was extensively used to provide pannage for swine, and the Normans also introduced the rabbit; both were probably important factors in preventing regeneration (the growth of saplings to replace older trees as they died off).

The Cistercian monks had a considerable influence on the ecological history of the uplands of northern England. The monasteries—Hexham, Fountains and Furness, for example—owned large tracts of hill country and the monks were efficient estate managers. They were sheep farmers on an enormous scale and large areas of woodland were cleared to provide grazing. The pollen record of many sites in upland Britain shows a decline in the quantity of tree pollen and an increase in the moorland and grassland species in the twelfth and thirteenth centuries.

In Ireland the elm decline occurred at about the same time as in England and was similarly followed by scattered temporary clearances. A major phase of disforestation has been dated at A.D. 300. It has been correlated with the beginning of the Christian era, the Celtic monasteries providing a stimulus to agriculture.

In Scotland, although some of the pine forest of the Highlands was destroyed by fire during the Viking raids, the main disforestation came much later than in England. Nevertheless, exploitation of timber in the sixteenth century, the expansion of sheep farming in the eighteenth and then the further cutting of timber in the First and Second World Wars have reduced the former 'Great Wood of Caledon' to a series of isolated remnants.

Oakwoods

By about 1200 the woodlands of lowland England began to have a scarcity value and to be managed for profit. The usual method was coppicing: a limited number of oak-trees, and sometimes also elms, would be allowed to remain as 'standards', while a lower layer of ash *(Fraxinus excelsior)*, maple *(Acer campestre)* and hazel *(Corylus avellana)* were regularly cut or 'coppiced'—a fourteen-year rotational cycle being customary. The fully grown standard trees provided the larger timber, while the coppice was used as a source of firewood, poles and charcoal. Thus an entirely artificial 'coppice-with-standards' woodland structure developed (see plate 5). In many places coppicing was abandoned during the nineteenth century, but locally it persisted up to the First World War. Often, although a wood may not have been cut for over fifty years, evidence of the former management of the woodland may be seen in the cluster of four or five slender trunks growing from a single, much thicker, stool.

At Hayley Wood in Cambridgeshire, in the early 1960s, coppicing was re-established by the local Naturalists' Trust on an experimental basis, so that the effects on the ground vegetation (the field or herb layer) could be studied. After cutting, the disturbance of the soil allows the establishment of open-ground plants such as the marsh thistle *(Cirsium palustre)*, and the amount of light reaching the ground encourages vigorous growth of species such as violets *(Viola)* and bluebell *(Endymion non-scriptus)*. As the coppice grows these plants eventually decline.

Woods may be approximately dated by the plants they contain. A wood that has grown from the abandonment of an area of scrub within the last generation or two will contain even-aged trees and there may be a considerable amount of ivy *(Hedera helix)* on the ground. On clay soils primroses *(Primula vulgaris)* may be common. A mature oakwood, even if it has been managed, is likely to have a different assemblage of plants, including dog's mercury *(Mercurialis perennis)* and perhaps, in East Anglian woodlands, the oxlip *(Primula elatior)*.

The common or stalked oak *(Quercus robur)* is the species which dominates the woodlands of the clay soils of the lowlands, while the sessile or durmast oak *(Q. petraea)* is characteristic of the more acid soils and rocky areas of the west and north of England, Wales, in some of the glens of the Scottish Highlands and here and there in Ireland (e.g. at Killarney). Mosses, liverworts, ferns and lichens are abundant in the damp environments of western oakwoods—at Keskadale Oaks, in the Lake District, there are some fifty species of mosses and liverworts. (Although similar in many respects to the other upland oakwoods, Wistman's Wood, on Dartmoor, page 31, consists of common oaks).

Beechwoods

Beechwoods (the origin of which is mentioned on page 17), grow on the steep slopes of the Chiltern Hills, the North and South Downs, the Cotswolds and locally in East Anglia on chalk and limestone soils; sometimes the woods on the chalk valley sides are known as 'hangers' (see plate 6). Holly *(Ilex aquifolium)* and yew *(Taxus baccata)* sometimes constitute a second tree layer, as does another evergreen, box *(Buxus sempervirens)*. Often, however, a shrub layer is absent from beechwoods. The herb layer may contain the following: wild strawberry *(Fragaria vesca)*, wood anemone *(Anemone nemorosa)* and the much rarer green helleborine *(Helleborus viridis)*.

Although beech is usually associated with chalky soils, it is also found in sandy areas, e.g. Burnham Beeches, Buckinghamshire, and Epping Forest. The soils are infertile, and so sandy beechwoods have a rather impoverished flora; the herb layer may be absent over quite large areas although large cushions of mosses, including *Leucobryum glaucum,* are common, and there may be patches of heather and ling in some of the more open spaces.

Scottish pinewood and birchwood

Tracts of native pine forest (plate 8) survive at Rothiemurchus Forest in Inverness-shire and on Deeside, Aberdeenshire. Sometimes the occasional rowan *(Sorbus aucuparia)* may occur amongst the pine, but pine stands are usually very pure. Taller shrubs are virtually absent, although the undershrub bilberry *(Vaccinium myrtillus)* may cover much of the ground and ferns are quite frequent. Heather grows where there is enough light. Amongst the rare plant species is the twinflower *(Linnaea borealis)*, with delicate, pink, funnel-like flowers. Many pinewoods are not regenerating because of sheep-grazing and damage done to the younger trees by deer. In many localities a long period of degeneration of the pinewood has resulted in the formation of a 'parkland' community in which scattered pines grow above an open expanse of heather and juniper *(Juniperus communis)*.

Birch-trees (plate 7) may grow in groups close to stands of pines in Scotland, often occupying the larger gaps in the pinewoods. They grow on a wide range of soils and are found up to 2,000 feet (600 metres) in the mountains of Perthshire. Grasses predominate in the ground layer vegetation of these northern birchwood communities and thus they are frequently used for grazing.

Plantations

Most of the Forestry Commission plantations which have been established since 1919, as well as many of those privately owned, are of conifers, as these produce marketable timber in a shorter time than deciduous trees such as the oak. The Commission owns about 750,000 acres (*c.* 300,000 hectares) in England, of which about 575,000 acres (*c.* 230,000 hectares) have been planted. The native Scots pine makes up substantial acreages, but more recently the Corsican pine *(Pinus nigra)* has been found more productive on some of the light sandy soils of eastern England, e.g. the Suffolk Sandlings. In upland areas other exotic species, such as the North American lodge-pole pine *(Pinus contorta)* and the Sitka spruce *(Picea sitchensis)*, have been found to grow well.

The landscape of the forestry plantation is very distinctive; the trees are arranged in rows with military exactness and the stands are usually kept free of underwood. Firebreaks—strips of land devoid of trees—are regularly cleared. Un-interrupted stands of a single species of conifer are very monotonous and so the Commission has frequently planted corridors of a variety of species, including native deciduous trees, along roads; this has greatly improved the appearance of, for example, Thetford Forest in Norfolk. The Forestry Commission has also placed emphasis on the use of forested areas for recreation, particularly those scenic areas that have been designated Forest Parks such as the Border Forest Park in Northumberland and Roxburghshire, the Forest of Dean, and the Queen Elizabeth park in Perthshire; picnic areas, camp-sites and car parks have been provided.

Here and there in the south and east of England some pine trees have established themselves in sandy heathland areas from nearby plantation sources; such communities are described as ' sub-spontaneous ' woodlands.

The ground beneath conifers, whether in natural pinewoods, plantations or sub-spontaneous growth, becomes covered in ' needles '. Fungi are important in coniferous woodland eco-systems, as they assist in the decomposition of these, many decomposer organisms being unable to consume the waxy needle material.

In the autumn, fungi such as the bitter *Clitocybe flaccida* and many species of *Boletus,* some of which are good to eat, are abundant in pinewoods and plantations. (Do not eat fungi without consulting a specialist work on the subject or on the advice of someone, such as a botanist, able to identify the edible species. A few woodland types are very poisonous.)

Woodland animal communities

A woodland ecosystem has a fairly well-defined physical structure; the canopy (the branches of the trees with their leaves), the shrub layer, the herb or ground layer, the top-soil, and the subsoil. The microclimate (daily temperature variation and humidity, for example) and light conditions vary greatly between the high canopy and ground level. Each layer of the forest plant community has its characteristic animals: earthworms may burrow several feet below the soil surface; woodlice are found in the litter of dead leaves and wood fragments on the forest floor; mice and voles operate at ground level; and squirrels search for food and build their nests (dreys) among the branches. This vertical zonation is also evident in the heights at which various species of bird make their nests: pheasants on the ground, warblers in the low tangle of bramble, the mistle-thrush on a branch some 12 to 20 feet (4 to 7 metres) from the ground, with rookeries in the high canopy. There are, however, many links between the different levels; birds nesting close to the ground may search for food amongst the branches, and caterpillars of the winter moth (*Operophtera brumata*) feed on tree-leaves but crawl down the tree-trunks to pupate (i.e. turn into a chrysalis) in the ground.

Oakwoods constitute perhaps the most complex communities in the British Isles—one estimate has it that there are over 220 insect species dependent on the oak. There are some fifty that make oak-galls or eat the tissues of oak-galls—small, often marble-like swellings on oak twigs and leaves. The ' oak-apple ', for example, is made by the larva of a gall-wasp (*Biorhiza pallida*); this in its turn may be parasitised by small creatures called Chalcids, but there may be other organisms living in the gall without actually harming the gall-former. Galls, therefore, constitute quite complex ecosystems in themselves.

Caterpillars of some eighty species of moth eat oak-leaves. Other leaf-feeders are aphids; these in turn provide food for other insects such as ladybirds. A hole in the trunk may provide a nesting site for a pair of woodpeckers, a home for squirrels or a colony of hornets. An individual tree may be the abode of tens of thousands of organisms.

The commonest small mammals in woodland are the bank vole and the wood mouse (*Apodemus sylvaticus*); these feed mainly on seeds and fruits such as acorns, haws and beech mast, although they may take the occasional beetle or caterpillar. They in their turn provide food for the tawny owl, the optimum population of which seems to be about thirty pairs to 1000 acres (about 400 hectares). It has been shown by research at Wytham Woods near Oxford that the population densities of small mammals and owls are closely related to one another (see page 6).

A patch of woodland may thus represent a historical monument of immense antiquity; it is also a living system of great complexity, consisting of countless food-chains with intricate ecological links between them.

3. GRASSLANDS

The range of grassland types

As woodland vegetation once covered most of Britain it follows that grasslands must usually be artificial communities. If not regularly grazed or mown they are invaded by scrub—often hawthorn—and eventually revert to woodland. There is, however, a continuous gradation between pastures produced as the result of sowing a blend of selected grass seeds, heavily stocked and regularly treated with lime and fertilisers, and those grassland areas subjected to very light grazing where little effort has been made to improve them. There is evidence that some grasslands might indeed benefit from heavier grazing. The degree of management or the extent of grazing pressure (whether by sheep and cattle or animals such as the rabbit) might thus form one basis for the classification of grasslands. Very often, however, classification is according to the character of the soil, and thus acid, neutral and basic grasslands are distinguished.

Acid grasslands

There are often transitions between acid grasslands and heathland or moorland communities. Bent-fescue grassland, dominated by species of *Agrostis* and *Festuca,* is characteristic of well-drained, acid, but not too poor, soils in uplands and is extensively used for grazing. Grass heaths, with soft grass *(Holcus mollis)* and wavy hair grass *(Deschampsia flexuosa)*, as well as bent and fescue, occur on sandy commons—heavy grazing tends to eliminate characteristic heathland species such as ling. Ill-drained upland areas often support a mat-grass community; these areas, dominated by *Nardus stricta,* are

23

monotonous, with few other species, and are of little value for grazing.

Neutral grasslands

These are to be found on lowland clays, loams and alluvial soils, and are usually subjected to much heavier mowing, trampling and grazing than acid grasslands. They may also receive considerable dressings of fertilisers. Amongst the grass species favoured by the farmer are timothy *(Phleum pratense)*, the meadow-grasses *(Poa pratensis* and *P. trivialis)* and cocksfoot *(Dactylis glomerata)*. White clover *(Trifolium repens)* is a particularly valuable plant because of its nitrogen-fixing properties. Other plants include the buttercups—field *(Ranunculus acris)* and bulbous *(R. bulbosus)*—which are found in damp meadows and are avoided by stock, being mildly poisonous, and ribwort plantain *(Plantago lanceolata)*, the pollen of which is one of the causes of hay fever.

Chalk and limestone grasslands (basic)

The grasslands of chalk downland of southern England have a very characteristic flora. The semi-natural grasslands of the North and South Downs, Salisbury Plain, Dorset and smaller areas in eastern England—on the ancient earthworks, Devil's Dyke and Fleam Dyke in Cambridgeshire for example —are of this type. The grasses include the fescues, where grazing is heavy, and erect brome *(Bromus erectus)* where it is not so severe. Other characteristic species include the stem-less thistle *(Cirsium acaulon)*, which has a rosette of very prickly leaves with a single dark crimson flower in the centre, and purging flax *(Linum catharticum)* with its small white bell-shaped flowers on thread-like stems. The pasque-flower *(Anemone pulsatilla)*, a rich combination of violet and gold and perhaps the most attractive of England's wild flowers, occurs in chalkland areas in Hertfordshire and sometimes also on other limestones. A number of species of orchid are also found on chalk and limestone soils, including some of the mountain limestones of Yorkshire.

Because of economic changes sheep are no longer pastured on some downland areas as extensively as formerly, and since 1954 rabbit grazing has been reduced. Some grassland areas are therefore now being invaded by hawthorn and juniper scrub.

Grassland animal communities

The disappearance of the rabbit following myxomatosis also had a profound effect on the animal ecology of grassland areas. The reduction in grazing gradually changed the floral composition of some grasslands. Grasses are eaten by the caterpillars of a number of butterflies, the small and Essex skippers *(Adopoea sylvestris and A. lineola)* for example. When, in 1955, the grasses were allowed to grow without disturbance instead of being eaten by the rabbits, the numbers of skippers increased. In the longer term, however, scrub invasion led to the local extinction of some grassland insect species.

An Oxfordshire study revealed the differences between the insects of short, highly grazed turf and long, almost ungrazed grassland. For example the commonest grasshopper in the area of rabbit-grazed turf was *Myrmeleotettix maculatus* while *Chorthippus parallelus* predominated in the longer grass.

Soil type also has a bearing on the character of the animal communities; thus molluscs (i.e. snails) tend to be common in chalky areas because of the availability of lime for their shells. The occurrence of certain species of plant on basic soils naturally influences the distributions of members of food-chains dependent upon them. Thus the chalk-hill blue butterfly *(Lysandra coridon)*, the caterpillars of which feed on several downland vetches, is a very characteristic butterfly of chalk grasslands.

The vertical zonation in grassland animal communities is less conspicuous than that seen in woodland, but exists nevertheless. For example, orb-web spiders spin their webs amongst the taller plants of a meadow, while large numbers—perhaps as many as 200 per square metre in the summer—of very small species of spider, such as *Hahnia nava,* live deep in the turf.

4. FIELD AND HEDGEROW

Arable lands

The forests that covered Britain in prehistoric times were complex communities with many species of animals and plants. The introduction of agriculture has removed the intricate network of interlocking relationships and substituted a highly simplified system. The complex food-webs of the forest have been replaced by communities in which a very small number of food-chains predominate; for example:

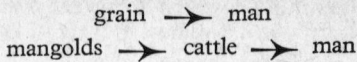

$$\text{grain} \longrightarrow \text{man}$$
$$\text{mangolds} \longrightarrow \text{cattle} \longrightarrow \text{man}$$

Almost every agricultural practice may be seen as the 'fattening up' of those parts of the farmland ecosystem which provide food (or raw materials) for mankind's use—crops and live-stock—and the prevention of 'side-tracking' by competing organisms. Thus weeds which might compete for mineral nutrients are removed by weeding or the use of herbicides; fungicides are used to control fungal diseases, and other chemicals are mobilised to kill insect, mite and eel-worm pests. Birds such as pigeons, which might compete with man for the fully grown crop or eat growing seedlings, are shot or are scared away, using a range of devices from the traditional scarecrow to the modern exploding detonator. Parasitic organisms that reduce the quality or quantity of products from animals are similarly controlled by such procedures as sheep-dipping and calf-vaccination.

In spite of the use of a range of herbicides, such as MCPA and 2,4-D, and a range of insecticides, croplands still represent quite varied communities; the corn poppy (Papaver rhoeas) is still occasionally to be seen in the English cornfield and the wild oat (Avena fatua) has proved difficult to eliminate. Many plants and animals may temporarily invade cropland from the surrounding hedgerows and rough ground. Frit-flies (Oscinella frit) may enter fields of oats from nearby unmanaged grassy areas and increase in numbers rapidly. Seven-spot ladybirds (Coccinella septempunctata) also some-times breed in cereals, where they live on aphids (greenfly). Probably over a hundred species of insect could be found in a typical arable field in summer.

Several birds, including the appropriately named corn-bunting (Emberiza calandra), which is to be found in the cereal, clover and lucerne fields of lowland England, find arable countryside an attractive habitat. The saying 'Good farming and partridges go hand in hand' has some ecological justification. Partridges (Perdix perdix) are truly birds of farmland, particularly light-soiled corn-growing country, but they require some rough areas—heathlands, hedgerows or gorse scrub—for nesting sites and cover. On the other hand they are seldom numerous on heath or moorland except where it adjoins cultivated land. The partridge forms an ecological link between the arable farmland and uncultivated ground communities taking food from both of them. A detailed pre-war study of the stomach contents of several hundred partridges from all over Britain showed a striking seasonal pattern in their feeding habits. In September nearly eighty per cent of their food was grain from stubble left after harvest. In October and November much land was ploughed, grain was scarcer and was supplemented by weed seeds and grass, the proportion of grass increasing through the winter and into the spring. In summer flowers and buds appeared in the diet, along with weed seeds and about ten per cent animal food (insects, etc.). Partridges formerly represented an additional crop from farmland—the largest bag of partridges ever recorded as being shot in a single day was at Holkham, Norfolk; there, on 7th November 1905, 1,671 birds were shot by eight guns. Changing patterns of farming, in particular the removal of hedgerows, together with disease, have substantially reduced the partridge population in the last three decades and it is unlikely that the coveys will ever again be as numerous as in the past.

Hedgerows

Hedges have been part of the English landscape for at least a thousand years; some are mentioned as boundaries in Anglo-Saxon charters. For many centuries piecemeal private enclosure of land continued. (Enclosure is technically the surrounding of land, usually common land, by fences or hedges, thereby making it private property.) It was possible in the thirteenth century to obtain a licence to enclose from the sovereign, but enclosure was on the whole discouraged—there were no fewer than eleven acts of Parliament between 1489 and 1624 prohibiting unauthorised enclosure.

Many hedges are no more than two hundred years old, and came into being as the result of the Enclosure Acts between 1750 and 1850. In Huntingdonshire and Northamptonshire, for example, more than half the total area was enclosed under these acts.

Hawthorn (*Crataegus monogyna*) is the commonest hedge-row species in most of eastern England and southern Scotland; this species grows well on a variety of soil types and because of its prickles, soon forms an effective barrier against stock; it was widely used by the enclosing landlords of the nineteenth century. Locally many other species have been employed; hedges of beech occur in north Somerset and in parts of East Anglia, blackthorn (*Prunus spinosa*) is used in Purbeck.

It has been suggested that one may ascertain the approximate age of a hedge by counting the number of species of shrub in it; if one multiplies the number in a thirty-yard stretch of hedgerow by a hundred, it is argued, one has the age of the boundary in years. The method is far from exact, there being variations according to the type of soil and part of the country, and so on, but the technique does give some indication of age. A hedge-bank with a dozen species is quite possibly a boundary dating from before the Norman conquest; one with a single species is unlikely to antedate the Enclosure Acts.

Hedges have recently come under serious attack; thus in a group of three parishes in Huntingdonshire there was a reduction between 1946 and 1965 from 71 to 20 miles of hedgerow (114 to 32 kilometres). Another study in the same part of the country revealed a loss of twelve per cent in the total length of hedge between 1965 and 1969.

Hedges are certainly costly to maintain; in 1965 it was estimated that £1,500 per annum in maintenance costs would be saved by clearing the 48 miles (75 kilometres) from a 2,000 acre (c. 800 hectare) holding. Enlargement of fields allows the use of sophisticated agricultural machinery and reduces labour requirements. The removal of hedges may also increase the area available for cultivation by some three per cent. There are some farmers who believe that the shading of crops around the edges of fields reduces yield and that hedgerows act as refuges for birds and animals that damage crops.

The problem is by no means simple. Hedges in England and Wales represent an important reserve of growing timber; oak, elm and ash trees, in particular, grow along the hedges of lowland areas. A Forestry Commission survey undertaken in the 1950s estimated that there were some 73 million park and hedgerow trees in the country. As shown above, many links exist between the open-field ecosystem and the surrounding hedgerow community; hedges may harbour pests but they also act as refuges for predators that feed on the pests of the farmland. (It has thus been argued that if fewer hedges were removed, less would have to be spent on pest control.) They may act as shelter-belts, reducing soil erosion; they provide cover for a subsidiary crop of game, and act as nature reserves, linking larger patches of woodland, copses and areas of scrub. Rare plants may survive in them and they have great aesthetic value; there are many who admire the appearance of a broken, slightly irregular landscape of hedged farmland with groups of elms silhouetted against the sky-line.

5. HEATH, MOOR AND MOUNTAIN

Heathland

Lowland heaths, dominated by common heather or ling sometimes accompanied by bell-heather *(Erica cinerea)* and bracken *(Pteridium aquilinum)* as well as shrubs such as gorse *(Ulex europaeus)* and broom *(Sarothamnus scoparius)*, are almost always on infertile, sandy soils. They appear in Hampshire, Surrey, the Weald of Kent and Sussex, Dorset (Thomas Hardy's Egdon Heath) and East Anglia—the Breckland of west Norfolk and the east Suffolk Sandlings.

In water-logged hollows in most of these areas (in some of the depressions in the heathland round Poole Harbour, for example) the cross-leaved heath *(Erica tetralix)* occurs along with species such as the sundew and bog moss. Such communities are sometimes described as wet heaths, but there is no sharp distinction between them and boglands (page 45).

Although there is limited evidence of some earlier forest clearance, it is probable that it was in the Neolithic (New Stone Age, *c.* 3,000—1,700 B.C. that most of the heaths of lowland England formed. Studies of the pollen and other plant remains from lake deposits in Norfolk, for example, have revealed that a striking change in Breckland, from mixed oak forest to a more open vegetation of grasses, bracken and heather, coincided with the beginnings of Neolithic culture in the area.

Once formed, heathlands were 'kept open' and the invasion of trees and shrubs was prevented, partly by controlled burning and partly by grazing—by cattle and ponies in some places, the open areas of the New Forest for example, and sheep elsewhere. Rabbits were also very important; there were eight warrens, each in an area of several thousand acres of heath specially managed for rabbit production in Breckland in the late eighteenth century. The earliest known documentary reference to rabbit warrening in the area is from 1300.

Heathland is thus a community depending for its continued existence upon a particular pattern of land use. Warrening as a full-time occupation disappeared several decades ago, and the rabbit population was almost wiped out by myxomatosis in 1954. Numbers of sheep in heathland areas have also declined. In the nineteenth and earlier centuries many heaths were burnt in a carefully controlled manner to discourage the growth of scrub and provide fresh, young heather shoots for stock, but this is now seldom practised.

Many species of plants and animals are dependent on this delicate balance between the natural processes of succession and the tendency of man's activities to maintain an open community free from trees. Heathland reptiles such as the adder *(Vipera berus)* and common lizard *(Lacerta vivipara)* are well camouflaged against a background of light grey-brown heather and bracken fragments. So too are the grayling butterfly *(Eumenis semele)* and the nightjar *(Caprimulgus europaeus)*, a ground-nesting bird whose numbers are declining in almost all heathland areas.

The Dartford warbler *(Sylvia undata)*, a small grey-brown and chestnut bird of the Hampshire and Dorset heaths, requires a vegetation of heather with a few isolated gorse bushes as singing and look-out perches. If an area is devoid of gorse or becomes too overgrown the warblers desert it. There is a moth, *Euzophera neophares,* that appears intermittently on the East Anglian heaths; the caterpillar of this species feeds on the fungus *Daldinia concentrica,* which seldom grows except on burnt gorse and birch twigs. Here is an example of a whole food-chain that would seem to be largely dependent on man's occasional interference in the heathland ecosystem.

Not only are heaths now becoming overgrown with birch and pine through the disappearance of grazing animals and the cessation of regular burning—the occasional uncontrolled conflagration caused by people throwing away lighted cigarette ends or abandoning picnic fires without extinguishing them has quite different ecological effects—but they are also rapidly being taken over for other uses. During the agricultural depression after the First World War large areas were acquired by the Forestry Commission. Some 50,000 acres (20,200 hectares) of the Breckland area of Norfolk are now beneath plantations, mostly of conifers. About 8,000 acres (3,200 hectares), much of it former heathland, are now forested in east Suffolk. Other heaths have been found to be worth reclaiming for arable when modern agricultural techniques are used. With the application of lime, and sometimes trace elements (very small quantities of substances such as boron and copper which are necessary for satisfactory plant growth), the use of deep ploughing and occasionally of summer irrigation, reasonable crops may be obtained. Other areas have been utilised as golf courses or for sand and gravel working. Thus in the 150 years following 1811 the area of heathland in Dorset was reduced from 75,000 acres to 25,000 acres (about 30,000 to 10,000 hectares); in east Suffolk 19,000 acres was reduced to 8,400 acres (7,700 to 3,400 hectares) between 1889 and 1966.

Moorland

Heather moorland is found on the hills formed from the resistant rocks of western and northern Britain, usually at altitudes of 1,000 to 2,000 feet (305 to 610 metres). Thus it covers much of Dartmoor and Bodmin Moor in south-west England, the Pennines and Cheviots, the North Yorkshire Moors, much of the Southern Uplands and the Highlands of Scotland, and Central Wales as well as much of Ireland. The plant community resembles that of the lowland heaths in being dominated by heather, but often other species of shrubs occur, such as crowberry *(Empetrum nigrum)*, with black fruits, cowberry *(Vaccinium vitis-idaea)*, with rather bitter red berries, and bilberry with delicious bluish-purple fruits. Lichens are also often abundant.

Soils are acid and peaty, but vary greatly in depth and dampness. There are seldom great thicknesses of peat on the sandy ridges left by the glaciers of the Ice Age in the valleys of the Scottish Highlands, but where gradients are gentle and the rocks below impermeable, as on the granite of Dartmoor, very thick layers of blanket peat may accumulate, particularly where rainfall totals are high (plate 10). Around the head of the East Dart the peat reaches 15 feet (about 5 metres) in thickness; it is almost as thick on some of the gritstone hills of the Pennines and in some of the Irish upland areas, e.g. the Sperrin Mountains in the counties of Londonderry and Tyrone and western Mayo. Bog moss, cotton-grass and purple moor grass are common constituents of blanket peat communities, their partially decomposed remains, along with fibrous material from the heather, making up much of the peat.

It has already been stated (pages 16-18) that many of these desolate heather-clad areas supported a forest cover for some four or five thousand years following the amelioration of the climate after the Ice Age, and that it was only after its removal that the peat began to accumulate and the moorland community developed. In places isolated fragments of woodland remain—Wistman's Wood, for example, a tangle of short, gnarled oak-trees festooned with ferns, amongst the granite boulders at 1,400 feet (*c.* 425 metres) near Two Bridges on

Dartmoor. The most important evidence, however, is provided by pollen grains and other plant remains preserved in the peat. It is possible that Mesolithic man (about 5,000 B.C.) started to clear the oak woodland from Dartmoor, but it was in the Neolithic period and Iron Age that disforestation went ahead most rapidly. It was probably completed by the tin-smelters who cleared most of the remaining woodland for charcoal in the twelfth and thirteenth centuries. Of course, the peaty layer is very thin on the upland areas that have been most recently cleared.

In parts of the southern Pennines, on Kinder Scout and Bleaklow for example, on the North Yorkshire Moors and on Dartmoor, the peatlands have suffered serious erosion in recent decades; here and there deep scars have been cut into the peat, occasional areas being entirely stripped. Opinions differ as to the reasons for this, some maintaining that the change is climatically induced, an increase in rainfall perhaps, others suggesting that severe burning and overgrazing are to blame.

Mountain communities

In the higher mountains a distinct altitudinal zonation of communities can usually be distinguished. For instance, the lower slopes of the Cairngorms carry remnants of birch and pine woodland. Above these are areas where heather is dominant and higher still is a zone in which crowberry and bilberry are the most conspicuous plants. The summits support a sparse, open vegetation of the moss *Rhaconitrium,* sometimes with the sedge *Carex bigelowii.*

In these high mountains plants are exposed to strong winds for long periods, and experience low temperatures. Snowfall may be heavy and the snow may remain for many months; thus on the summit of Ben Nevis (4,406 feet, 1,343 metres) snow first appears in September and may lie continuously from October until May. The growing season is therefore short.

The low structure and tufted habit characteristic of 'alpine-arctic' plants gives protection from the severe winds of high altitudes. Mosses and lichens are particularly resistant to such extreme conditions and are generally abundant in high mountain vegetation, sometimes occurring along with arctic species such as dwarf birch *(Betula nana).* A number of high-altitude plant species have a 'viviparous' habit; vegetative buds replace flowers, and on becoming detached take root in the soil. This is a useful adaptation to environments where the extreme

1. *Relationships between organisms: the hole of a green woodpecker (Picus viridis) in a birch (Betula pendula) beneath the fungus Polyporus betulinus.*

2. Members of a food-chain: a barn owl (*Tyto alba*) with a vole.

3. *Adaptation to environment: the snipe (Gallinago gallinago) is concealed by both its camouflaged plumage and its habit of hiding its conspicuous bill amongst its feathers.*

4. *An example of a hydrosere: a freshwater pool being gradually invaded by reedswamp and woodland.*

5. *Man's management of woodland: mature oaks in Surrey with an understorey of coppice—trees regularly cut back for firewood, etc.*

6. *A 'hanger' of beech in Sussex: beechwoods commonly grow on steep slopes in chalk and limestone country.*

7. *A Perthshire birchwood: grasses predominate in the ground layer vegetation.*

8. *A pine forest in Perthshire, showing some regeneration; stands of pine are usually very pure, with few tall shrubs.*

9. *Mountain pasture in Caernarvonshire, with scree and scrub oak covering the slopes.*

10. *Blanket peat on Dartmoor, now showing signs of erosion; thick layers accumulate in areas of high rainfall and with impermeable rock below.*

11. *Marram-grass is usually the dominant plant in sand-dunes, providing almost the only food supply and shelter for animals.*

12. *The Brecon Beacons are one of ten national parks in England and Wales designated under the National Parks and Access to the Countryside Act (1949).*

13. *The Devil's Dyke, a chalk grassland community in Cambridgeshire that has remained undisturbed for generations. Part of it is now leased as a nature reserve by the local naturalists' trust.*

shortness of the season precludes the setting of seed.

Grouse and deer

Many of the upland areas of Britain are managed so as to produce a crop of grouse *(Lagopus lagopus scoticus)*, sheep, or red deer *(Cervus elaphus)*. On many of these moors the shepherds and gamekeepers have traditionally burned stands of old, leggy heather in order to increase the amount of younger growth. Ideally a rotation is practised so that all the heather is renewed every eleven to fifteen years.

Heather shoots, flowers and seedheads are by far the most important food of grouse, supplemented by the fruits of other moorland shrubs. Cock grouse establish territories for themselves on the moors, defending them against rivals; the possession of a territory appears to be essential for breeding. In most years more young grouse are produced than territories are available, and when one is vacated it is usually soon reoccupied.

Herds of red deer graze the higher slopes of many mountain areas in Scotland, moving on to the lower ground in winter and occasionally venturing into woodland (where they may do considerable damage) in severe weather. Scottish deer have few natural enemies, although a recent study on the Isle of Rhum revealed that thirteen per cent of dead calves examined had injuries inflicted by eagles, and if not culled numbers soon increase to such an extent that serious overgrazing may result.

Both the grouse and red deer are well adapted in food requirements and behaviour to their upland environment, and it seems that careful management of the heather moors of the British uplands should enable a substantial crop of both species to be taken without the population being seriously threatened.

6. FRESHWATER AND WETLAND

The process by which a stretch of open water is gradually transformed first into marshland or bog and ultimately into a woodland community has already been described (pages 11-12) and ecological succession will not be discussed further here. In this section the animals and plants of a range of freshwater and wetland habitats are described as they might appear at a single point in time, without reference to the long-term changes that may be occurring within them.

Still waters

Stagnant and near-stagnant water-bodies are common habitats in lowland Britain. There are hundreds of miles of 'lodes' or drainage ditches in Fenland, and farm and village ponds are almost ubiquitous. Abandoned gravel and clay pits offer a similar habitat, as do the backwaters of larger rivers and streams.

Even a very small pool offers a variety of environments: an outer margin of emergent vegetation, a zone of submerged and floating vegetation, the surface-film, open water and bottom mud.

The edge of the pool is likely to be inhabited by plants rooted in the mud beneath the water but which protrude above it; examples include the bulrush *(Scirpus lacustris)* and arrowhead *(Sagittaria sagittifolia)*. These plants constitute an important link between the water and terrestrial and aerial environments. They provide food for animals such as water voles *(Arvicola amphibius)* and also a routeway into and out of the water for insects such as dragonflies and damsel-flies that spend only part of their life in freshwater.

Water-lilies are rooted in the bottom of the pond, but have large floating leaves which reduce the amount of light penetrating the water, although the lower surfaces of lily pads provide sites for egg deposition for some animals.

The plants of the region of submerged vegetation may include several species of pondweeds *(Potamogeton)* and the Canadian pondweed *(Elodea canadensis)*, which was introduced to Britain in the nineteenth century and spread very quickly as small fragments grow rapidly into new plants. The weeds provide shelter, oxygen (through photosynthesis) and, directly or indirectly, a source of food for a wide variety of species of pond snails *(Limnaea)*, beetles, water-mites, dragonfly nymphs, leeches and flatworms. Some of these animals feed on others, and an elaborate network of food-chains exists.

The surface film—the boundary between the air and water —constitutes a habitat with plenty of oxygen and an abundance of food in the form of small animals that have accidentally fallen on the water. Several creatures show a remarkable degree of adaptation to this highly specialised environment. Examples are the pond-skater *(Gerris)* and water-measurer *(Hydrometra stagnorum)* which, having long slender legs, spread their light weight over a large area. Hanging from the underside of the film are the larvae of insects such as mosquitoes. Duckweeds *(Lemna)* are plants that live on the surface, their roots hanging free in the water.

A whole series of complex food-chains are based on the micro-organisms or plankton living in the open water of a pond. Planktonic algae include *Volvox* which is occasionally abundant enough to colour the water green. Animals include rotifers and water-fleas. These are eaten by small fish which in their turn are eaten by predacious fish such as the pike *(Esox lucius)* and some water birds.

The remains of dead plants and animals accumulate on the pond bottom and thus the mud may be exceedingly rich in decaying organic matter—a source of food for large numbers of organisms. Light is absent, as may be oxygen. Creatures such as ' bloodworms '—the larvae of midges—and some true worms, such as *Tubifex*, live in the mud with only the tail protruding into the water above; these possess haemoglobin, which absorbs oxygen and enables the animals to survive when very little oxygen is present.

Running water

Conditions are very different in a stony stream; a strong current prevents the growth of large masses of vegetation and could injure or carry away small animals without structures enabling them to attach themselves to stones. The nymph (larva) of the mayfly *(Ecdyonurus)* is very flat in shape, and thus is able to keep close to the stones and so avoid the strongest currents. Oxygen, however, is much more abundant than in still waters, for air is continually being dissolved at the stream's turbulent surface. Caddis-fly larvae, protected by cases that they build up from fragments of shell or sand, may be present, together with stonefly larvae and flatworms, these creatures live partly on organic detritus that falls into the stream. Trout *(Salmo trutta)* live in the deeper pools in the becks and burns of upland Britain, while smaller fish, such as the miller's thumb *(Cottus gobio)*, dart amongst the stones.

Downstream deposition of silt and clay particles occurs, particularly on the insides of bends in the river, and plants such as water crowfoot *(Ranunculus aquatilis)* may colonise these little stretches of mud. Weed grows in patches and provides shelter for coarse fish such as the minnow *(Phoxinus phoxinus)*, gudgeon *(Gobio gobio)* and roach *(Rutilus rutilus)* as well as many much smaller creatures such as the green hydra *(Chlorohydra viridissima)*, easily overlooked because of its diminutive size, about ¼ inch or 0.7 centimetres when extended, and its colour which blends in with the plants to which it is usually attached.

As the river nears the sea it becomes tidal and 'brackish'—the salt content increasing. The hydroid *Cordylophora lacustris* is found in estuaries along with a number of snails such as Jenkin's spire shell *(Hydrobia)*, a species which prior to the 1890s was confined to brackish waters, but has now extended its range to freshwater habitats. It appears to be parthenogenetic, reproducing without a male sex.

Fens

Fen vegetation is to be found on alkaline peat soils. Although their area has been greatly reduced by drainage and reclamation, such communities are not uncommon in lowland England. They surround the Norfolk Broads, and isolated relicts in a predominantly agricultural landscape are to be found at Wicken in Cambridgeshire and Wood Walton in Huntingdonshire. Similar habitats occur in the northern and western districts of the country but are rarer. Leighton Moss, Lancashire, close to the shores of Morecambe Bay, is in some respects similar to the localities listed for eastern England.

Continued waterlogging of lakeside and riverside areas impedes the decomposition of plant material so that peat accumulates. In the most characteristic localities the water comes from areas of calcium-rich rocks such as chalk and limestone, the lime imposing an alkaline reaction on the Fen peat. The dominant plant may be the reed but fens are often quite rich in species, some of which have brightly coloured flowers, so that some fenland areas in summer present a most attractive picture; examples include meadow-sweet *(Filipendula ulmaria)*, yellow flag *(Iris pseudacorus)* and marsh orchid *(Dactylorchis incarnata)*. Both reed and sedge *(Cladium mariscus)* are used for thatching and are regularly cut in the marshland areas of eastern England. Money from sales of thatching materials from selected portions of Wicken Fen helps to pay for the upkeep of the nature reserve as a whole. The cutting produces a number of ecological changes; heavy cutting of the sedge, for instance, results in the expansion of purple moor grass *(Molinia caerulea)*.

Probably because of the wide range of plant species, fens have a rich insect fauna; at Wicken over a thousand species of beetles and about 750 kinds of butterflies and moths have been noted. The specialities of the East Anglian fens include two butterflies, the swallow-tail *(Papilio machaon)* and the large copper *(Lycaena dispar)*, the food-plants of which are respectively milk parsley *(Peucedanum palustre)* and great water dock *(Rumex hydrolapathum)*. The large copper became extinct in England in 1851, the last specimens being taken at Bottisham near Cambridge, but a closely related continental form has recently been reintroduced.

Birds of the reed beds include the reed warbler *(Acrocephalus scirpaceus)*, an inconspicuous buff-brown bird that interweaves or suspends its nest from three or four growing reed stems. It is woven from strips of grass, reed and sedge and some-times lined with the downy flowers of the reed. Both bird and nest are well camouflaged.

In the 1930s the coypu *(Myocastor coypus)*, a large brown South American rodent, was introduced to eastern England on an experimental scale for fur-farming. A number escaped and by the 1960s there were established populations in many wetland areas in Norfolk and Suffolk and smaller numbers elsewhere. As the animals ate large quantities of reeds and damaged agricultural crops, a Government-backed campaign to reduce them was organised. It has substantially reduced numbers.

Boglands

The term 'bog' is often used to describe any area so saturated with water that one tends to sink into it. Ecologically the term is reserved for plant communities dominated by bog moss *(Sphagnum)* and in which the peaty soil is acid in reaction, in contrast to the alkaline fens.

Valley bogs occur in depressions in the heaths of southern England, e.g. Denny Bog in the New Forest and on Studland Heath in Dorset. As well as heathland species there may be rushes *(Juncus)*, sedges *(Carex)* and occasional patches of cotton-grass *(Eriophorum)*.

In northern England and Scotland the term 'moss' is sometimes used, e.g. Malham Tarn Moss, near Settle, York-shire. These communities are usually raised bogs, so called as they have slightly domed surfaces and are often slightly higher than the surrounding land. Sometimes, underneath the acid *Sphagnum* peat is a layer of fen peat, frequently with lake deposits lower down. This is the case at Malham. Peat accumulation is able to continue above the height of the stream entering a depression, the plants being nourished by nutrients dissolved in the rainwater. These bogs are therefore usually in areas of high rainfall and are very acid. Because of the paucity of plant nutrients, areas of this sort are poor in plant species; however cotton-grass, with its fluffy white heads, and bog asphodel *(Narthecium ossifragum)*, with bright yellow flowers, may break the monotony of large expanses of hummocky *Sphagnum*.

One plant that has an interesting adaptation to the nutrient-lacking bogland ecosystem is the sundew *(Drosera)*—a species with a rosette of round flat leaves, each leaf being edged by sticky red hairs which trap insects. Bladderwort *(Utricularia)* is an example of a plant that is sometimes found in bog pools and that overcomes the problem of nitrogen deficiency in a similar way—it has sophisticated underwater suction traps to catch insects. These two plants are able to obtain the nitrogen they require from the proteins in the tissues of the creatures they capture.

Many bogs in England have been drained, but a spectacular example exists at Tregaron in central Wales and they are also common in Ireland.

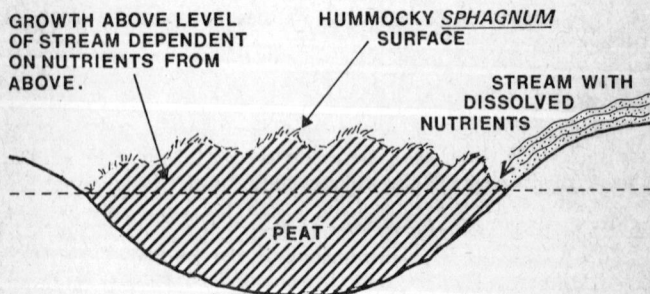

Fig. 5: Cross-section of raised bog.

7. COASTAL COMMUNITIES

Cliffs and islands

Cliff ecosystems have connections with both terrestrial and marine environments. The rocky surface may be affected by salt spray and some food-chains lead from the sea towards the land; thus sea-birds such as kittiwakes *(Rissa tridactyla)* or guillemots *(Uria aalge)* that nest in large numbers on cliffs (e.g. below Scarborough Castle and at Flamborough Head in Yorkshire) take food from the sea, yet their young may be taken by rats.

On small islands, such as Skokholm and Skomer off Pembrokeshire, some of the Hebrides, the Farne Islands off Northumberland or Rathlin off the north coast of Ireland, the interconnections between marine and terrestrial ecological processes may be readily perceived. For example, a character-istic and widely distributed cliff plant is thrift or sea-pink *(Armeria maritima);* it tends, however, to be overwhelmed by grasses, such as the fescues, in the absence of grazing. On many islands rabbits graze on the ledges of quite steep cliffs and so maintain an environment favourable to this attractive plant. Rabbit burrows provide breeding places for puffins *(Fratercula arctica)* and on some islands, such as Rhum and Skokholm, also for Manx shearwaters *(Procellaria puffinus)*. Near large sea-bird colonies the droppings stimulate plant growth—Inner Farne has a rich flora of flowering plants —but too many bird droppings may poison the soil. Also associated with such nesting colonies are a number of lower plants such as the alga *Prasiola crispa,* a plant with a very wide distribution. It has been found amongst breeding sea-birds on Bear Island in the Arctic as well as in the Antarctic penguin rookeries and is evidently very well adapted to its highly specialised niche. Another common cliff-plant that may be encouraged by bird droppings is the bright yellow lichen *Xanthoria.*

Sea shores

Many people have probably had their interest in natural history and ecology aroused by peering into a rocky pool left on the shore by the retreating tide. The hermit crab *(Eupagurus bernhardus),* which shelters its soft, unprotected abdomen in a whelk-shell *(Buccinum undatum),* is usually found to be sharing its borrowed home with the worm *Nereis fucata*. Sometimes a sea-anemone *(Calliactis parasitica)*, and a hydroid *(Hydractinea echinata)*, grow on top of the shell, providing a vivid demonstration of the network of ecological relationships that exists in any community. The shell may have been abandoned, but most of these relationships are commensal. Simpler linkages, however, also exist in the rock-pool ecosystem: sea-anemones feed on tiny fish and shrimps, and limpets *(Patella)* consume the algae that grow on the sides of the pool.

Because of the twice daily rise and fall of the tides and because of differences in the amount of exposure sea-shore animals and plants can tolerate, organisms are often distributed in parallel zones along the foreshore. For example, in the lowermost zone, which is seldom exposed to the air, will be found the yard-long ribbons of brown oakweed *(Laminaria)*. A little higher the rocks revealed at low water are covered with slippery masses of serrated wrack *(Fucus serratus)* with jagged edges to its fronds and above that, the bladder-wrack *(F. vesiculosus)* which is distinguished by air-bladders arranged in pairs on either side of a midrib. In the splash zone, which is never entirely below water, the rocks are covered by the smooth black lichen *Verrucaria*.

The character of the beach material also greatly influences the types of organisms found. On sandy shores seaweeds are not able to anchor themselves and many of the animals have an elongate form, an adaptation to a burrowing way of life, e.g. the razor-shell *(Ensis)*, the sand eel *(Ammodytes lancea)* and numerous worms. On less exposed, muddier shores, where there is more organic material in the substratum, coiled castings on the surface betray to the fisherman seeking bait the presence of lugworms *(Arenicola)* in their tubes below.

Shingle beaches often do not show many signs of life. Yet even here, along the strand line—the zone along which decaying seaweed and other material thrown up by the sea is deposited—flies and beetles as well as scavengers such as sand-hoppers *(Talitrus)* may be active. And higher up the beach, along shingle ridges seldom reached by the tide, clumps of the conspicuous yellow horned poppy *(Glaucium flavum)* may be found.

Salt-marshes

The plants found most frequently in salt-marsh environments are halophytes, plants that can withstand very high accumulations of salt in the soil. They are characteristic of low, gently shelving and sheltered coasts, particularly along estuaries, where marine deposition is supplemented by material brought in by streams from the land. Different levels of the salt-marsh are affected by the sea to different degrees—plants of the lower part may be awash at high tide—and so a zonation often exists. Samphire *(Salicornia)* grows on the mud close to the sea, while higher up may be found plant communities with sea-lavender *(Limonium vulgare)* and sea-purslane *(Halimione portulacoides);* stands of sea-rush *(Juncus maritimus)*, mark the transition to a terrestrial environment.

Salt-marshes are communities which are continually changing. The deposition of silt and the accumulation of dead vegetation result in a rise in the level of the marsh, but at the same time erosion by the sea may be occurring. The result is often a maze of muddy creeks, salt pools, islets—a no-man's-land between land and sea, characteristic of stretches of the north Norfolk coast and the mouths of a number of the rivers of Essex and Suffolk, such as the Stour, Deben and Alde.

Sand-dunes

Sand-dunes accumulate on shingle spits and low islands; splendid examples of dune ridges exist on the Snook, a long peninsula on Holy Island, Northumberland, and on Scolt Head Island, Norfolk. The successional stages in dune development have already been described (page 13) and so the plant communities need not be further detailed here.

The dominant plant is almost always marram-grass. A number of insects feed on it—for example the caterpillars of the shore wainscot moth *(Leucania litoralis)*. These larvae feed on the marram leaves at night, seeking concealment by burrowing into the sand during daylight.

The marram tussocks, the clusters of roots beneath them and the organic debris around them provide the only shelter and, directly or indirectly, the only source of food in the austere dune habitat. On a sunny summer day the sand surface may get very hot and the air above it extremely dry; some small animals exposed to such conditions would quickly lose moisture. In the clumps of grass will be found snails, plant-eating beetles and woodlice. Yet animal life is not entirely confined to the tussocks and a number of creatures venture on to the bare sand, even by day. Amongst these are predators such as spiders, tiger beetles *(Cicindela hybrida)* and hunting wasps. The commonest bird among the sand-dunes is usually the meadow pipit *(Anthus pratensis)*. It feeds on insects, spiders, caterpillars and seeds and makes its nest in the tufts of marram. Its young may be found by a marauding stoat—and so a food-chain is completed.

8. THE EFFECTS OF MAN

Certainly since the Neolithic, and probably since the Mesolithic period, man has had an appreciable effect on the plant and animal communities of the British Isles. Forests have either been cleared or profoundly modified, grasslands and heathlands formed and wetlands have been drained. Ecosystems of all kinds have been manipulated to yield the food and raw materials required by man. Sometimes these changes have had unfortunate results, not only for the animal and plant life, but for man himself.

Soil erosion

The soil is obviously a fundamental element in any terrestrial ecosystem. It is the medium in which plants grow, the reservoir of most plant foods and the environment in which many of the bacteria and fungi responsible for the recycling of nutrients are active. About five per cent of an average soil is organic matter. The removal of the natural vegetation cover renders the soil liable to erosion by wind and water and when soils are damaged or completely removed the consequences are serious.

Perhaps the most spectacular examples of soil erosion occurred in the 1930s in the Great Plains of the United States. Some have estimated that a quarter of the farmlands of the plains were damaged; in one storm, on 11th May 1934, about 300 million tons of topsoil are estimated to have been removed. Between 1934 and 1938 there were 263 dust-storms in Texas and Oklahoma alone. They occurred because the plains sometimes experience periods of several years' drought alternating with groups of more normal, moister years. The indigenous grasses of the prairies were adapted to this regime, cereal crops are not. In the drier years autumn-sown wheat failed to germinate and so there was nothing to hold the loose soils. The strong winds blew the topsoil from thousands of square miles, causing the immense dust-storms.

The limestone ridges of southern France and Spain once

supported a near-continuous cover of forest and scrub vegetation, but a long history of clearance and grazing, particularly by goats, has enabled storm rains to remove much of the soil. Today only in isolated depressions do the red-brown Mediterranean soils remain and the uplands are sun-scorched wildernesses.

Ecological damage on this scale has not occurred in Britain, but soil erosion on a more subdued scale certainly exists. By the early 1970s substantial areas of the Cambridgeshire fens that were shown on maps of a century earlier as covered by thick peat were reduced to a ' skirt ' soil—a soil in which the upper layer contains much organic matter but with an infertile sandy layer below. Some of the peat had decomposed, oxidised by continual ploughing, but some had been removed by the wind. By the year 2000, it is calculated, a further 24 inches (61 centimetres) of topsoil will have been removed and only small patches of continuous fen peat will remain.

Elsewhere in eastern England removal of the hedges has resulted in appreciable soil erosion. It is a common sight, during a dry spring, to see piles of blown soil accumulating by the roadside in the sandy area of east Suffolk.

Peat erosion in moorland areas has already been mentioned (page 32).

Introductions

Man's activities in introducing organisms to an area, whether deliberately or accidentally, can sometimes be almost as devastating as his removal of them. The introduction of Canadian pondweed and the coypu have already been noted (pages 42 and 45).

The American grey squirrel *(Sciurus carolinensis)* was introduced to about thirty places in Britain between 1876 and 1920. Eighty years after its successful establishment it had occupied some 30,000 square miles. It is now found in almost every English and Welsh county and in a number of areas in Scotland and Ireland. Not only has this species displaced the native red squirrel *(Sciurus vulgaris)*, but it is a serious forest pest, damaging hardwood trees by stripping the bark.

Great devastation was caused by the arrival of the Asian chestnut blight *(Endothia parasitica)* in eastern North America in 1911. The American chestnut tree *(Castanea dentata)* had no resistance to the disease, and the infection spread very rapidly through the eastern states. By 1950 most of the chestnut trees east of the Mississippi river were dead, and the ecology of the American forests was much altered.

Analogous changes have occurred in lacustrine and riverine habitats. Lampreys *(Petromyzon marinus)* have long been established in Lake Ontario in eastern North America, but their progress inland was prevented by the Niagara Falls. In 1829 the Welland Canal was constructed to by-pass the falls and in due course the lamprey entered Lake Erie. It got as far as Lake Huron and Lake Michigan by 1937 and reached Lake Superior in 1946. The lamprey population exploded and the stocks of grey trout *(Salvelinus namaycush)* in the waters of the Great Lakes declined catastrophically, as lampreys attach themselves to fish and eat them alive. The commercial fisheries of the region were ruined and the whole ecological community of the inland waters was transformed.

Extinction of animals

Through hunting and the destruction of their habitats by man, some dozens of species of animals and birds have been eliminated in comparatively recent times and many others have been brought to the verge of extinction.

The dodo was exterminated at the end of the seventeenth century; sailors visiting the island of Mauritius in the Indian Ocean found the bird both easy to kill and good to eat. The same fate befell the great auk *(Alca impennis)*, a penguin-like bird that once nested in the Faeroes, Newfoundland and possibly also in Ireland. The last great auks ever seen may have been the pair swimming in Belfast Lough on 23rd September 1845.

Perhaps the most spectacular story of the extinction of a species is that of the passenger pigeon of North America. A single flock of this species was estimated to number over a thousand million and when they migrated south in the autumn, the pigeons passed in immense clouds, filling the skies for three days on end. They broke branches with their weight when they settled on trees and were netted and shot by the hundred thousand. Initially this hunting seems to have had little effect on numbers, but eventually the colossal slaughter, together with the reduction of the passenger pigeon's forest habitats brought about its extinction. The last specimen died in a Mid-Western zoo in 1914.

The American bison or buffalo *(Bison bison)* was almost exterminated. At the opening of the nineteenth century about sixty million buffalo roamed the American prairies, providing the chief means of support of the Indian population. By 1889 only 541 survived in the USA. They were killed for meat (especially to feed the workers on the railroads) and hides and for political reasons (in attempts to starve and suppress the Indians). One traveller, in 1847, described how he came upon a vast area of carcasses, shot and left to decompose, only the marketable tongues having been removed. Under protection the population has since recovered.

By the early years of the present century these American birds and mammals had been exterminated: Carolina parakeet, passenger pigeon, heath hen, Labrador duck, Arizona elk and giant sea mink.

When colonists arrived in southern Africa they found vast herds of elephants and other game. The animals were killed, not just for food (and ivory), but in order to clear the veld for use as farmland. The bluebuck had been exterminated by 1800. The only elephants left in Cape Province today are a herd of forty-four in the 20,000 acre (8,090 hectare) Addo Elephant Park near Port Elizabeth, where they live in a semi-wild environment. They are regularly fed with fruit.

Several animals, whilst they survive elsewhere, were rendered extinct in Britain within the historic period. The brown bear lingered in Scotland until about the tenth century, and the wolf was also formerly widespread in the British Isles. It became extinct in England in the sixteenth century and in Scotland and Ireland in the eighteenth. The beaver *(Castor fiber)* continued to live in the valley of the Teifi in Cardiganshire until the end of the twelfth century.

Place-names such as Boarshill in Oxfordshire, Boarhunt in Hampshire and Wild Boar Fell in the Lake District testify to the former abundance of the wild-boar *(Sus scrofa)*. A twelfth-century writer described the countryside near London as being infested with boar. The animal was hunted mercilessly, however, and the last wild-boar in England was seen in Westmorland about 1680.

Pollution

Pollution may be defined as loss of purity; so strictly speaking, any contamination of air, water or soil by artificial means may be said to pollute them, but usually the term is applied when the contamination has appreciable biological effect. Anything that is not normally present in an environment, and that affects organisms when added to it, may therefore be regarded as causing pollution.

Sewage, for example, when it flows into rivers or the sea not only causes an unpleasant smell and may act as a vehicle for the spread of disease but may also greatly alter the ecological equilibrium. Micro-organisms increase in numbers, utilising the organic material as a source of food. Eventually, however, such matter will entirely decompose.

This is not the case with many synthetic, chemical pollutants; the organo-chloride pesticides, for example, leave residues that are not naturally decomposed, i.e. are not bio-degradable. Also they may be stored in the tissues of organisms almost indefinitely—in the fat of mammals and birds for example. One result of this is that pesticide residues tend to accumulate towards the end of food-chains. In California in 1949, 1954 and 1957 very low concentrations (one part in fifty million parts of water) of a pesticide were applied to a lake in an attempt to destroy the midges that made sport fishing in the area miserable. The chemical was concentrated 265 times in the plankton, 500 times in small fish and up to 85,000 in the predacious fish and fish-eating birds.

In England an abrupt decline was noted in the population of the peregrine falcon *(Falco peregrinus)* following the widespread availability of certain types of insecticide in the 1950s. In 1961 only 31 per cent of the peregrine territories in southern England were occupied, and young were reared in no more than four per cent. Analyses of peregrine corpses showed that their tissues contained appreciable amounts of organo-chloride residues. Some of the eggs which remained unhatched after a long period of incubation were similarly contaminated. Peregrines eat pigeons, and pigeons are known to take in quite large amounts of pesticides from the farm-lands on which they feed. In another study, the unhatched eggs of an insect-eating bird, the red-backed shrike *(Lanius collurio)*, were found to contain quantities of a pesticide called chlordane.

Because ecosystems consist of such a complex network of ecological relationships, the application of pesticides may have quite different effects from those intended. In an experiment in Huntingdonshire DDT was used to control the caterpillars of the cabbage white butterfly *(Pieris rapae)* on Brussels sprouts. After the initial effect of the spraying, however, the survival of the pest improved. It was found that the natural enemies of the caterpillars, ground beetles for example, that normally lived in the soil and climbed up the stems at night in search of caterpillars, were being killed in large numbers.

Pollution of the sea

In March 1967 the oil tanker *Torrey Canyon* crashed into the Seven Stones reef, about 20 miles due west of Land's End, at a speed of 17 knots, releasing thousands of tons of crude oil into the sea. This was not an isolated incident, but a particularly spectacular instance from amongst the many spillages that have affected the world's coastlines. It was intensively studied and much information on the ecological effects of contamination was collected.

Slicks of oil on water may resemble patches of seaweed or shoals of fish, and sea-birds in search of food are attracted. The oiling of birds' feathers interferes with their movement and if badly oiled they eventually starve. Birds may also swallow substantial quantities of oil and become poisoned. Following the *Torrey Canyon* catastrophe 7,900 sea-birds were treated at cleansing centres, but only 443 were eventually released and probably few of these survived.

The problem is a continuing one: on 11th January 1971 two oil tankers, the *Paracas* and *Texaco Caribbean* collided and sank in the English Channel. The following day the *Brandenberg* crashed into the wrecks before warning buoys had been placed and she, too, was lost. In April 1971 severe oil pollution occurred in the Orkneys, and many thousands of birds were killed. In March of that year a special study of 925 miles of beach around the British Isles resulted in the finding of 374 dead and twenty living oiled sea-birds. The puffin population seems to have been particularly badly hit in recent years.

Following the contamination of the beaches of Devon and Cornwall by *Torrey Canyon* oil, thousands of gallons of detergent were sprayed in an attempt to disperse it. Unfortunately, these detergents are a good deal more poisonous than crude oil to many organisms. Mussels, oysters, limpets and shrimps are all quickly killed in quite low concentrations of emulsifier.

The initial pollution of the French coast, in Brittany, was even more serious than that of Cornwall. There was nearly 2 feet (60 centimetres) of oil on the beach at Perros-Guirec. Perhaps because of concern for their inshore fishing, the French adopted a rather different approach to the problem of the contamination of the beaches by oil. Sawdust, chalk and ash were distributed on the oil at sea, the object being to agglomerate the oil rather than to disperse it with emulsifiers. Some of it sank to the sea-floor; as much as possible of what reached the beaches was removed and dumped in old quarries inland. This technique seems to have been less damaging to marine life than spraying.

Sometimes substances even more dangerous than crude oil pollute the sea. In January 1972 a freighter foundered off the Cornish coast; its cargo included 286 drums of sodium cyanide, a very poisonous chemical, 274 drums of tolulene diviso-cyanate, which gives off a poisonous vapour, 290 drums of volatile ethyl acetate and a further 402 drums of a proprietory chemical which was both poisonous and inflammable. Many were never recovered.

The examples of marine contamination mentioned above were accidental. But increasingly chemicals are being deliberately dumped at sea. It recently came to light that a Swedish firm was regularly dumping chlorinated hydrocarbons into the sea off northern Norway. This triggered off intensive investigations and many Norwegian fishermen reported finding unlabelled canisters and drums in their nets. These containers —some of them leaking—were found to contain poisons. In the summer of 1969 a research vessel cruising in the North Sea between Scotland and Denmark encountered a belt, several miles in length, of dead fish and plankton, presumably killed by these or similar toxic chemicals.

River pollution

Freshwater organisms are often extremely susceptible to pollutants. In one series of experiments, batches of young tadpoles sustained a mortality of about seventy per cent in a concentration of ten parts per million of DDT. In solutions of one part per million twenty per cent perished. The development of those which survived was retarded. DDT is amongst the commonest pollutants of our rivers, streams and ponds: it may be blown by the wind from nearby farmland, or carried by birds, or contamination of water bodies may arise from the common practice of abandoning empty pesticide containers in a nearby stream or drainage ditch when a spraying programme has been completed.

But there are many other activities that despoil rivers. The Trent is perhaps the most polluted river in Britain: it receives lead, zinc and copper compounds, cyanide, detergents and coal-dust from the steelworks, collieries, chemical and rubber works and literally dozens of local-authority sewage disposal undertakings along its length. Fish are unable to live in long stretches of the river Trent or in some of its tributaries.

The Rhine is the only river in western Europe filthier than the Trent. In the autumn of 1970 a shipping firm was accused of dumping 20,000 tons of poisonous factory waste in the Rhine. The lawyers acting for those responsible pleaded that tipping such a 'mere thimbleful' could have had only a negligible effect on the river's pollution problem!

Another form of pollution of rivers and streams results from the widespread use of agricultural fertilisers. Much of the nitrogenous fertiliser added to farmland is washed out into land-drains and eventually reaches rivers and lakes; this sometimes results in a burst of biological activity, a 'bloom' of algae. The supplies of oxygen dissolved in the water are used up and fish and other forms of life die.

A rather different type of interference with the normal life processes in freshwater is thermal pollution—the artificial heating of the waters. This normally occurs as the result of the release of cooling water from power stations into a river—there are eighteen power stations along the river Trent. The rate of decay of organic material accelerates and so oxygen is also consumed at a faster rate. As well as this, fish tend to absorb poisons more rapidly in warm water than in cool. It has been found that coarse fish will not normally live in water above 88°F (31°C).

The picture is not, however, an entirely gloomy one; great attempts have been made, for example, to improve the quality of water in the Thames. In 1957 there was almost no oxygen dissolved in the waters of the Thames for a period of nine months along a stretch of about forty-five miles. Fish, except for eels, were absent. The sewage disposal plants along the river were modified and effluent levels in the river were strikingly reduced. Fish have returned—fifty-five species were recorded between 1967 and 1971.

Atmospheric pollution

The adverse effects of atmospheric pollution on human health are well known. There is a close correlation between the amount of smoke in the atmosphere and the incidence of bronchitis. Some authors have suggested that the four-day smog in December 1952 cost the lives of 4,000 Londoners.

Coal and fuel oils usually contain some sulphur (some oils have a content of over three per cent sulphur) and when these are burnt serious pollution of the atmosphere by sulphur dioxide may occur; this substance is harmful to many organisms, including man. Research close to many large urban and industrial areas, in Newcastle, Belfast and in South Wales, for example, has shown that lichens are almost totally excluded from an area close to the source of the pollution and that this 'lichen desert' is surrounded by a 'struggle zone' in which a limited number of species of lichen grow with difficulty; it has been established that a major cause of this is the presence of sulphur dioxide in the atmosphere.

Even more insidious may be the effects of pollution of the atmosphere by metals emitted during some industrial processes—mercury, arsenic, lead, copper and zinc. In November 1969 a cloud of arsenic fumes escaped from a chemical works in Gloucestershire and drifted across nearby farmland where it caused the death of four cattle and the sickness of many more.

But although factory chimneys and chemical works are an important cause of air pollution, motor-car exhaust can be equally serious. Cars emit a number of pollutants: oxides of nitrogen, unburnt hydrocarbons, lead and carbon monoxide. Lead is a cumulative poison—substantial quantities have been found in the body tissues of people who live and work in cities, and there is some evidence from experiments with animals that the hydrocarbon material could cause cancer.

Radioactivity

Following the nuclear weapon tests at Eniwetok atoll in the Pacific in 1948, genetic abnormalities were discovered in ten species of plants; and radioactivity may induce changes (mutations), the effects of which may not become apparent for generations.

While low dosages of radioactivity are used with great success in the treatment of the early stages of cancer, accidental exposure to severe radiation causes symptoms such as loss of hair, leukaemia and death.

The radioactive material released by a nuclear explosion or the disposal of industrial waste (such as that from nuclear power stations) may be absorbed by organisms and passed along a food-chain, as is the case with other pollutants. Following nuclear tests in the Arctic, a very high concentration of radioactive caesium was noticed in the lichens of the Alaskan tundra, and this was eventually acccumulated by the caribou *(Rangifer arcticus)* that fed on them.

The signing of the Moscow Treaty in 1963 banning the testing of nuclear weapons in the atmosphere, in space and in water has caused amounts of radioactive fall-out from the atmosphere to decline. But the problem of disposal of radioactive wastes and the risk of an accidental escape from a nuclear power station remains.

The world-wide problem

The problem of pollution is ubiquitous. Migrating animals and birds, ocean currents and winds carry pollutants throughout the world; appreciable quantities of organo-chloride residues have been identified in the bodies of seals and penguins in the Antarctic.

Attempts are being made to develop pesticides which decay after they have done their work; precautions against pollution become ever more stringent. But the use of DDT and similar chemicals is nevertheless increasing, particularly in the developing countries, some of which are short of food. As populations increase and living standards rise, more chemicals and metals will be required and the risk of pollution by occasional accidents will increase. With an ever rising demand for energy the risk of atmospheric pollution and thermal pollution from conventional power stations and the problem of disposal of effluent from nuclear power stations remain. (Nuclear power accounts for about six per cent of Britain's production.) Constant vigilance is necessary: one must remember that pollutants tend to accumulate in the end members of food-chains, and that at the end of a very large number of food-chains is man.

9. CONSERVATION

Conservation has been defined as the maintenance of a favourable balance between the human population and the resources on which it depends. It may demand that the rate of consumption of a resource be reduced, or it may involve a change in the method of exploitation, but in either case the intention is to benefit posterity.

Conservation is much more than mere protection or preservation; it usually requires the active management of an environment, so that it may be made to yield whatever is required of it by man—timber, water, fish or simply recreation—both in the short and in the longer term.

Some examples of conservation

Forests in upland regions check soil erosion, absorb the water of sudden rainstorms and so prevent flooding, provide a habitat for varied wildlife and are capable of withstanding heavy recreational pressure. This does not mean the conservationist is opposed to the exploitation of the timber resources of the forest. He would advocate a policy of selective felling and the replanting of trees in cleared areas; a management programme that ensures that the forest will yield timber for future generations is to be welcomed. It is the clear-felling of large areas, followed by practices such as heavy grazing which prevent recolonisation by forest species, that should be avoided.

Switzerland is an example of a country which has an extremely successful forest conservation policy. Laws were enacted in 1902 to ensure that the proportion of the country that was forest-covered was not reduced and prohibiting clear-felling—this followed a series of disastrous floods in the late nineteenth century which were blamed on the disforestation and exposure of steep upland slopes. The result is that Swiss forests produce nearly 130 million cubic feet of timber per annum and yet one quarter of the country is always under woodland. Locally, special 'shelterwoods' are maintained to protect villages from avalanches.

Monoculture, the growing of the same crop on an area of ground year after year, ultimately leads to soil exhaustion. Ploughing of farmland up and down the slope is apt to cause soil erosion by gullying. Crop-rotation, strip-cropping and contour-ploughing are soil conservation practices which maintain the long-term structure and fertility of the soil.

There are probably less than one thousand blue whales (*Balaenoptera musculus*) in the world's oceans; in 1938 there were estimated to be 100,000. The effects of intensive hunting of whales in the period following the Second World War—stimulated by a world shortage of fats—has resulted in several species of whale being threatened with extinction. The International Whaling Commission has established a quota —a maximum number of whales to be taken each season— and specified dates between which whales may not be caught. Reserves have been established; whaling has been banned in large areas of the ocean, and certain species may not be taken at all. Important as these conservation measures are it is doubtful whether anything less than a complete ban on whaling for several years can enable numbers to recover sufficiently for the long-term survival of the whaling industry to be assured.

In the absence of natural predators, populations of animals, whether greenfly or elephants, tend to increase. Ultimately the population density becomes so great that the food sources on which the animals depend are depleted. It is then likely that disease will become apparent and many individuals may starve. Thus when the Isle of Rhum in the Hebrides was acquired by the Nature Conservancy in March 1957, the moorland habitat had been severely over-grazed by sheep and red deer. A research programme was started in an attempt to determine the ideal population density and what cull should be made each year to maintain a healthy and stable deer population. It is hoped that the results of these studies will enable other Scottish deer forests to be managed so that substantial yields of venison may be obtained without the risk of degrading the habitat as a whole.

Britain's nature reserves and the Nature Conservancy

An important aspect of conservation is the preservation of areas of natural and near-natural vegetation for scientific study. It is essential that examples of unmodified ecosystems should continue to exist so that the way in which they function can be investigated. If man is to be able to exploit modified ecosystems satisfactorily ecologists must be able to compare their workings with those of systems that remain in a pristine state.

The Nature Conservancy was empowered, under the National Parks and Access to the Countryside Act (1949), to establish 'national nature reserves'. It was laid down in the act that the aim should be to provide special opportunities for the study and preservation of the flora, fauna, geological and physiographical features of Great Britain.

There are well over a hundred national nature reserves covering about a quarter of a million acres (c. 101,000 hectares). They include coastal sites such as Scolt Head Island on the Norfolk coast, heaths such as those at Studland in Dorset, Westleton in East Suffolk and Weeting in Norfolk, upland communities such as Moor House in the northern Pennines and Beinn Eighe in the Scottish Highlands, as well as the islands of St. Kilda and Rhum in the Hebrides.

There are woodlands at Monks Wood in Huntingdonshire and Yarner Wood in Devon.

The reserves vary in size from a few to tens of thousands of acres; some are owned by the Conservancy, others are leased from landowners. All are managed so as to maintain their ecological diversity and preserve the indigenous animals and plants. In some of these areas public access is allowed freely to all parts of the reserve but where the interests of conservation and research necessitate restrictions permits to visit must be sought.

Under Section 23 of the 1949 act the Nature Conservancy has the duty of designating 'sites of special scientific importance'; these are usually areas that are less important scientifically than the national nature reserves, but nevertheless interesting by reason of their 'flora, fauna or geological or physiographic features'. The designation of a site of special scientific importance—there are some two thousand of them—does not affect the ownership of land, but provides official recognition of the importance of the sites. The local planning authority has a legal duty to consult with the Nature Conservancy before development is allowed on a site designated in this way and in some instances the Conservancy has intervened for the protection of the areas.

The Conservancy was reconstituted and became part of the Natural Environment Research Council in June 1965. Its headquarters are at 19/20 Belgrave Square, London SW1.

Britain's national parks and the Countryside Commission

The first section of the 1949 act decreed that: 'There shall be a National Parks Commission which shall be charged with the duty of . . . the preservation and enhancement of natural beauty in England and Wales . . . particularly in areas designated under this Act as National Parks or as areas of outstanding natural beauty.' The Countryside Commission, as the National Parks Commission was later renamed, is therefore responsible for the conservation of landscapes. It was further charged with providing facilities for those visiting national parks for the purposes of open-air recreation and the study of nature.

The designation of an area as a national park does not imply any change in the ownership of land, nor, usually, is the pattern of land use affected to any great extent. It is realised that man's activities have interacted with natural processes for many generations in the development of the landscape—for example the dry-stone walls of the Yorkshire Dales and the Derbyshire Peak District add greatly to the interest of the hillsides they cross. Forestry, water catchment, agriculture and sometimes mining and quarrying are allowed to continue, although stringent planning controls on development are maintained.

Since 1949 the following national parks have been designated: Dartmoor, Exmoor, the Brecon Beacons, the Pembrokeshire Coast, Snowdonia, the Derbyshire Peak District, the Yorkshire Dales, the North Yorkshire Moors, the Lake District and North Northumberland (Cheviot Hills and Roman Wall).

Ideally, the traditional occupations of the region should continue in harmony with outdoor recreation and nature conservation, but this has not always been the case; the interests of the research ecologist, sheep farmer, water engineer, forester and rambler do not coincide. Also, in a few cases, major developments have been allowed in National Parks 'in the national interest'; examples include Fylingdales early warning station and a large potash mine on the North Yorkshire Moors and Trawsfynydd nuclear power station in North Wales.

'Areas of outstanding natural beauty' include landscapes that, while less spectacular than those of the national parks, are nevertheless regarded as being in need of protection and for which special planning procedures are relevant; examples include the Northumberland and Norfolk coasts and the North and South Downs.

In addition the Countryside Commission is responsible for long-distance footpaths, such as the Pennine Way, a route of some 250 miles from Edale in the Peak District to the Cheviot Hills in Northumberland.

Non-Government conservation organisations

The Nature Conservancy and the Countryside Commission are state-sponsored organisations; so too is the Forestry Commission. Important work in the field of nature conservation is also done by many independent organisations.

Perhaps the best known is the National Trust, a society founded in 1895 'to preserve places of natural beauty and historic interest'. It was incorporated by act of Parliament in 1907, but is not financed by the state. The Trust maintains a large number of country houses and has also been able to preserve substantial tracts of open country such as Wicken Sedge Fen in Cambridgeshire, one of the few remaining areas of wetland vegetation in the Fens, and Rough Tor, an area of Cornish moorland. The Trust has recently made great efforts to secure the remaining stretches of unspoilt coastal scenery and protects the dunes and pinewoods at Formby Point, Lancashire, and the magnificent cliff scenery at Tintagel and Boscastle in Cornwall. Similar trusts operate in Scotland and Northern Ireland.

The Royal Society for the Protection of Birds, appreciating that birds can only survive in the context of their natural habitats, maintains over forty nature reserves from the sea-bird colonies of Noss and Fetlar in Shetland to the coastal marshes of Minsmere and Havergate Island in Suffolk where rare marshland birds such as the avocet *(Recurvirostra avosetta)* and bittern *(Botaurus stellaris)* breed.

Almost every county in England has a county naturalists' or conservation trust; a total of five trusts cover Wales and the Scottish Wildlife Trust serves Scotland. These local trusts administer over five hundred reserves of local importance, operate nature trails and undertake other educational work. Some of the reserves administered by the trusts are quite small, e.g. disused gravel and chalk pits or stretches of abandoned railway line, but are nevertheless managed so as to preserve the variety and interest of their vegetation and wildlife. Others are more spectacular: the Cambridgeshire and Isle of Ely Naturalists' Trust has concluded a lease of a substantial length of Devil's Dyke, near Newmarket. The dyke is an ancient earthwork, dating from about A.D. 500, stretching many miles across the chalk from the edge of the Fens to the wooded clay country of Suffolk. It is one of the few places where a chalk grassland community has remained undisturbed for generations. In 1971 the British Petroleum Company gave funds to the Glamorgan County Trust to enable them to purchase Long Hole Cliff, one of the most splendid stretches of the rugged coastline of the Gower peninsula.

A recent venture into the conservation and outdoor recreation field has been the development of country parks by local authorities (with Countryside Commission encouragement). The object is to concentrate visitors at places where provision can be made for them in terms of car parks, toilets, picnic areas and nature trails, the theory being that pressure on other areas which would be spoilt by heavy recreational use will be relieved. One of the first of these (designated in 1969) was provided by Lancashire County Council at Beacon Fell, an isolated eminence 873 feet (267 metres) high within the Bowland Forest area of outstanding natural beauty. The fell commands panoramic views over the Ribble valley, the Lancashire plain and the Bowland Hills and makes an attractive destination for excursions from the urban areas of north-western England. The 271 acres (110 hectares) contain 115 acres of woodland and large acreages of rough pasture and moorland; it is intended that this range of habitats should be maintained and as far as possible the wildlife encouraged. At a similar park, Pow Hill, close to the boundary between Northumberland and Durham, visitors can enjoy walks over moorland country with vistas over a reservoir. Rustic picnic tables are provided and information boards explain something of the natural history of the area.

10. THE ROLE OF ECOLOGY IN A CROWDED WORLD

As human populations increase—and the world population is now (1973) over 3,700 million and increasing at the rate of about 180,000 every 24 hours—and as living standards rise, demands for food and raw materials increase. If reasonable supplies of food, water, timber and natural fibres (wool, cotton, sisal, jute etc.) are to continue to be available, the ecosystems which provide them—farmland, grassland, forest, river and ocean—must be exploited prudently. The working of ecosystems, both natural and artificial, must be well understood if yields of animal and plant material harvested from them are not to decline.

For example, in order that the yield of timber from Britain's forests and plantations shall be maintained at a high level, at the Forestry Commission Research Centre at Alice Holt, Farnham, Surrey, the whole of the forest ecosystem is studied. The relationships between climate, trees and soils are investigated, together with the effects of insects, birds, deer, squirrels and mice on tree growth.

Alternatives to pesticides

In view of the harmful effects of some pesticides, alternatives to their use are being sought. Various attempts have been made to exploit natural ecological relationships such as those between a predator and its prey, or a parasite and its host in an attempt to achieve biological control. The scale insect, accidentally introduced to California from Australia, was devastating the citrus fruit industry, so a ladybird was introduced to control the pest. An introduction in the opposite direction, from the western USA to Australia, resulted in the spread of prickly pear cactus *(Opuntia)* through Australia's pasture lands. Eventually a caterpillar was found that fed on the spiky plant, and the cactus was destroyed in many parts of its new range. The spectacular reduction of rabbit numbers by myxomatosis, in both Europe and Australia, provides another example of biological control.

Another technique is the use of sterile male insects to prevent breeding. In the USA vast numbers of screw-worm flies, the grub of which eats animals' living flesh, were bred, subjected to radiation in order to induce sterility, and released. The sterile males mated with normal wild females and, of course, the eggs eventually produced were infertile.

A major pest of many coniferous trees grown in British forestry plantations, especially Scots pine and spruce, is a fungus *Fomes annosus;* it causes internal decay of the living trees. An area that has not been used for growing timber for some time is generally free from infection, but the thinning of a stand of trees leaves cut stumps which are colonised by the airborne spores of the fungus. Colonisation is followed by the spread of infection down the main root system and thence to the roots of nearby living trees. In some of Britain's forests the level of infection may reach seventy per cent of all trees. After the main tree crop is felled the fungus remains in the cut tree stumps, and on replanting losses may be even higher. Although some reduction in the incidence of this pest has been obtained using chemicals, the most successful approach has been through biological control. Spores of a saprophytic fungus, *Peniophora gigantea,* are painted on to the stumps so that they decompose before *Fomes annosus* has a chance to establish itself.

Some aphid pests are attracted towards sheets of reflecting metal foil spread out in fields. The aphids tend normally to fly upwards towards the sky, but are confused by the patches of ' sky ' in the fields. The gatherings of aphids on the sheets can be dealt with without spreading insecticide over a wide area. A rather similar technique involves the use of synthetic pheremones, substances produced by female insects to attract males.

Sometimes, after a careful study of the ecology of a pest,

THE IMPORTANCE OF VARIETY

a number of these biological techniques, supplemented perhaps by judicious use of chemical methods, may be combined in an attempt to achieve integrated control.

The importance of variety

Natural ecosystems are highly complex; artificial communities—improved grasslands, pine plantations and wheatfields, for example—are much simplified: the organisms required by man are encouraged, the others eliminated. Often these simple systems are unstable. Thus the growth of cereals in the same fields year after year may result in soil erosion, and the cultivation of potatoes in the same soil for long periods results in the increase in the infestation levels of eel-worm pests. Single-species ecosystems such as block forestry plantations may similarly suffer from the rapid spread of pests.

There may, therefore, be advantages in imitating nature by introducing a measure of diversity. Hedgerows not only constitute windbreaks but provide a habitat for ladybirds which consume the aphid pests of farmland. Patches of rough ground may allow a crop of game-birds as well as one of barley to be taken from a farm; copses and small woodlands provide cover for wildlife as well as timber. Often there may thus be economic as well as aesthetic and ecological arguments for a farmland landscape including a 'mosaic' of plant communities.

Multiple use of land

In a country the size of Britain, inhabited by 56 million people, there is great competition for land. Some uses of land are mutually exclusive—obviously land used for urban development is no good for agriculture or forestry. But there may be circumstances in which land may be used for three or four purposes at once. There is no reason why an upland area should not be used for sheep grazing, water catchment and recreation—possibly also for military training. Deciding the most appropriate uses for land is often difficult and may require detailed ecological research. For example, studies in the Yorkshire hills by the Fylde Water Board suggest that when peat-covered moorland used for water catchment is planted with conifers there is a reduction in water yield. Some of the water is evaporated from the branches before it reaches the ground, and some of the water that does soak into the soil is taken up by the trees.

Land may be re-used; open-cast coal-mines can sometimes be quite speedily restored to agriculture and some abandoned mineral workings are suitable for conversion into wildfowl refuges, nature reserves or for angling and water sports. In the Darent valley an area of 266 acres (108 hectares) of gravel pits and surrounding land owned by the Kent Sand and Gravel Company was successfully turned into a reserve in which over a thousand pairs of birds, belonging to fifty-seven species, bred within ten years. Nesting sites were provided and woodland planted; the area now holds the most concentrated breeding population of duck in Kent.

There are thus many ways in which ecology has contributed appreciably to mankind's well-being. But particularly important, in an urbanised society, is the provision of areas of beauty and solitude, where man's interference is limited, for recreation and renewal. Not only the climber, rambler, angler, pony-trekker and wildfowler but also the artist and the lover of quiet places have an ally in the ecologist with his need for the conservation of a variety of ecosystems for scientific study.

GLOSSARY

Algae: a primitive group of flowerless plants.

Autecology: ecology of a single species or group of closely related organisms.

Biosphere: the part of the earth's crust, surface waters and atmosphere that is favourable to life.

Biotic factor: the influence of living organisms, in contrast to factors such as edaphic and climatic.

Bog: marsh-like community underlain by waterlogged, very acid peat.

Brackish: intermediate between salt and fresh water.

Calcicole: plant that grows in calcareous (lime-rich) soils.

Calcifuge: plant that grows in acid (lime-free) soils.

Canopy: upper part of a woodland ecosystem; the zone of tree branches.

Carbohydrate: energy-producing organic compound of carbon, hydrogen and water such as starch.

Carnivore: animal that feeds on other animals, e.g. spider, owl, stoat.

Chlorophyll: green colouring matter in plants necessary for photosynthesis.

Climax: end-point of a succession; a complex, relatively stable plant community that is in equilibrium with the climate and soil of a locality.

Commensalism: literally 'sharing the same table'; a food-sharing ecological relationship.

Community: in original ecological sense, a unit of vegetation that has recognisable characteristics distinguishing it from others, e.g. oakwood, heathland. Now sometimes an association of animals, or animals and plants.

Conservation: more than mere protection or preservation; the maintenance of a balance between population and resources.

Ecosystem: a segment of nature, including animals and plants plus their inorganic environment. The concept may be applied to a small unit such as a piece of rotting timber or a larger entity such as a woodland or island.

Edaphic factor: the influence of soil.

Element (chemical): a substance that cannot be broken down into simpler substances by chemical means, e.g. oxygen, nitrogen, hydrogen, iron, boron.

Fen: a lowland marsh-like vegetation growing over a water-logged neutral or alkali peat soil.

Food-chain: series of organisms that feed upon one another, thus:

oak-leaves → caterpillars → titmice → sparrow-hawk.

Food-web: network of interconnected food-chains.

Habitat: type of environment in which an organism normally lives.

Haemoglobin: red pigment that absorbs oxygen in the blood of certain animals.

Halophyte: plant adapted to live in salty soils.

Herbivore: plant-eating animal, e.g. aphid, rabbit, deer.

Herb layer or field layer: layer of plants covering the ground.

Hydrophyte: plant adapted to life in water.

Hydrosere: set of plant-communities that develops in wet environments, e.g. open water — reedswamp — carr — fen woodland.

Inorganic: not derived from plant or animal material.

Larva: caterpillar; one of the stages in the life-cycle of an insect.

Leguminous: legume or pod-bearing; leguminous plants, e.g. peas, vetches, clovers, often have nitrogen-fixing bacteria in nodules on their roots.

Microclimate: a modification of the local climate due to the immediate environment; thus the temperature and humidity (dampness) will be different beneath a tree from those in an open field.

Microhabitat: small-scale habitat; e.g. bird's nest, dung, animal carcass.

Nymph: aquatic larva of insect such as dragonfly.

Omnivore: animal that eats food of a wide variety of types.

Organic: derived from a living organism.

Organism: plant or animal.

Oxidation: chemical combination of a substance with oxygen, e.g. on burning.

Parasite: organism living within or upon another (the host) and obtaining part or all of its food from it.

Peat: partially decomposed vegetable material that has accumulated under waterlogged conditions.

Photosynthesis: building up by plants with chlorophyll of complex organic compounds from simple, inorganic substances, using the energy of sunlight.

Pollination: the transfer of pollen grains from the male to the female part of a flower or from one flower to another.

Pollution: loss of purity; contamination having appreciable biological effect.

Producer: organism capable of building up organic materials from inorganic sources (i.e. green plant).

Protein: complex organic compound of carbon, hydrogen, oxygen and nitrogen and sometimes sulphur and phosphorus. The essential nitrogen-containing food of animals.

Pupa: inactive stage in the life-cycle of an insect; chrysalis.

Raised bog: domed, acid peat community formed by the accumulation of plant material above the level at which water enters the depression in which the bog is situated.

Respiration: process by which organisms obtain energy, usually by the oxidation of material such as carbohydrates.

Saprophyte: plant, such as a fungus, that obtains its food from the dead remains of another organism.

Sere: set of communities that succeed one another in a particular environment.

Species: set of very similar individuals which are able to interbreed amongst themselves and produce fertile offspring.

Species-network: the whole system of ecological relationships that exist between the organisms in an area. The concept is wider than that of a food-web, as it includes ecological relationships other than food linkages—e.g. that between a flower and its pollinator—but narrower than an ecosystem, as inorganic materials—soils, air, water—are not included.

Succession: process of orderly community change; e.g. gradual development of scrub and then woodland on sand-dunes.

Symbiosis: the living together of two species of organisms in close association for mutual benefit; e.g. nitrogen-fixing bacteria and leguminous plants.

Synecology: ecology of a whole community or ecosystem; cf. autecology.

Valley bog: acid peat community in a depression in a lowland area.

Xerophyte: plant adapted to a dry environment.

BIBLIOGRAPHY

The books listed below vary in their approach; some are elementary, others are quite advanced text-books.

General ecological texts
A Guide to Field Biology; J. Sankey; Longmans, 1958.

Introduction to Field Biology; D. P. Bennet and D. A. Humphries; Arnold, 1965.

The above are 'how to do it' books, and explain how simple ecological surveys may be conducted.

Animal Ecology; C. S. Elton; Sidgwick and Jackson, 1927 —reprinted Methuen, 1966.

Britain's Green Mantle; A. G. Tansley; Allen and Unwin, 1949.

British Plant Life; W. B. Turrill; Collins, 1948.

Ecology; E. P. Odum; Holt, Rinehart and Winston, 1963.

Fundamentals of Ecology; E. P. Odum; Saunders, 1959.

The Pattern of Animal Communities; C. S. Elton; Methuen, 1966.

Books about specific habitats
Chalkland Ecology; J. Sankey; Heinemann, 1966.

Mountains and Moorlands; W. H. Pearsall; Collins, 1950 —reprinted as Fontana paperback.

The Observer's Book of Pond Life; J. Clegg; Warne, 1956.

The Sea Shore; C. M. Yonge; Collins, 1949—reprinted as Fontana paperback.

Woodland Ecology; E. G. Neal; Heinemann, 1958.

Conservation and pollution
Before Nature Dies; J. Dorst; Collins, 1970.

Pesticides and Pollution; K. Mellanby; Collins, 1967—reprinted as Fontana paperback.

INDEX OF PLACES

Printed by C. I. Thomas & Sons (Haverfordwest) Ltd., Press Buildings, Merlin's Bridge, Haverfordwest, Pembrokeshire.